No Day Without A Line

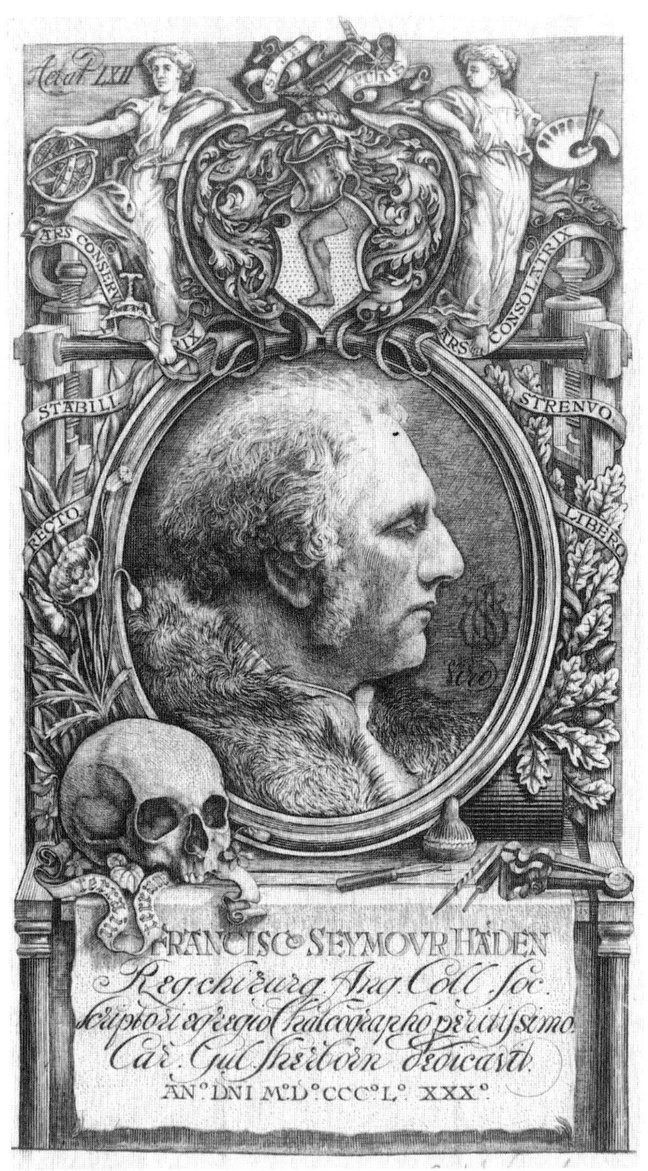

Charles William Sherborn (1831–1912)
RE65 *Portrait of Francis Seymour Haden* 1880
Engraving 149 x 86 mm

No Day Without A Line

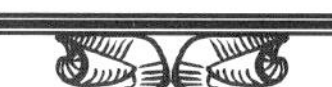

The History of
The Royal Society of Painter-Printmakers
1880–1999

by
Martin Hopkinson

with a list of the Diploma Collection by
Clare Tilbury

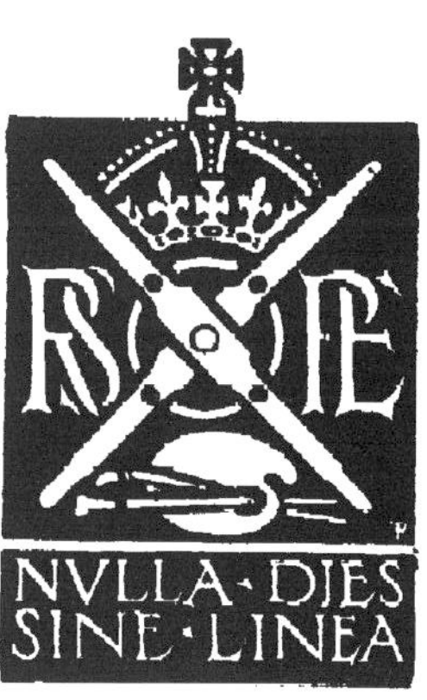

ASHMOLEAN MUSEUM OXFORD
in association with the Royal Society of Painter-Printmakers
1999

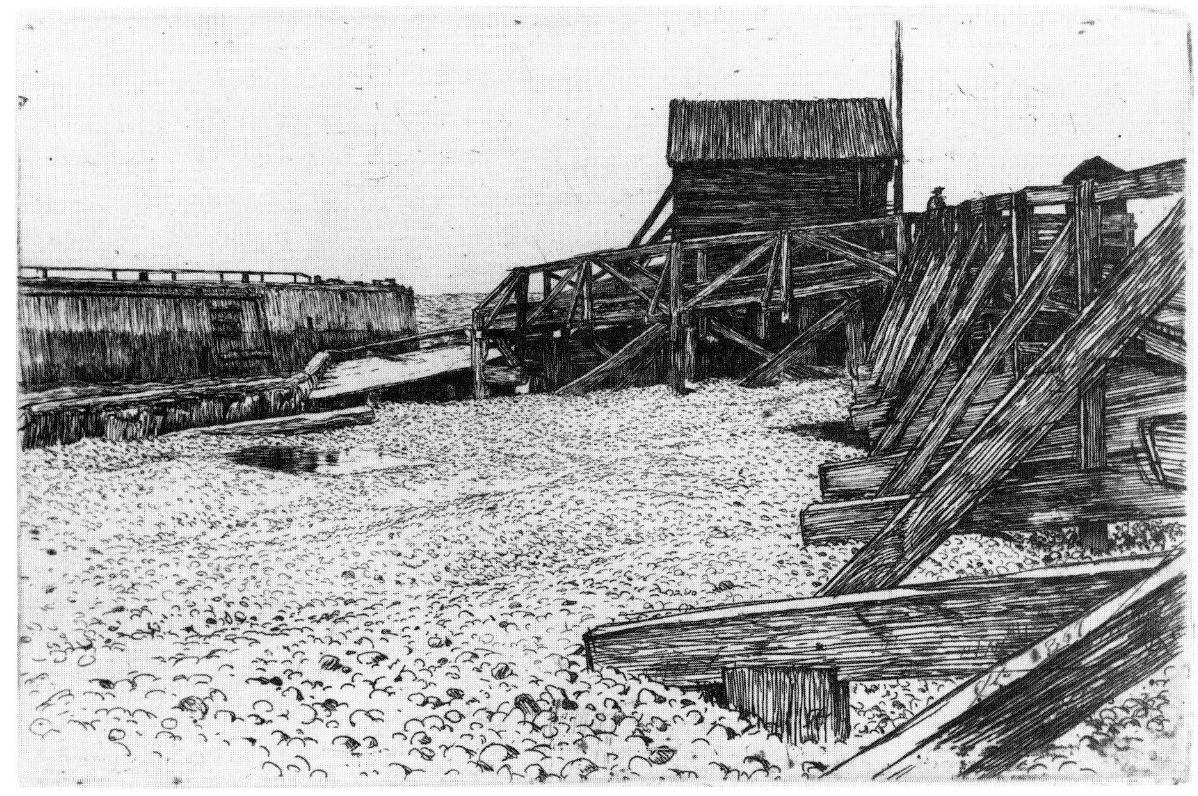

Charles Samuel Keene (1823-1891) RE12   *Walberswick Pier* 1867   Etching & drypoint 103 x 161 mm

Published, with a generous grant from the Paul Mellon Centre for British Art, to coincide with an exhibition of the Society's Diploma Works at the Ashmolean Museum.

ISBN 1 85444 119 1 paperback
      1 85444 096 9 hardback

British Library Cataloguing in Publications Data: a catalogue record of this book is available from the British Library.

Cover illustration: *Putting on Tights,* 1926, by Dame Laura Knight, D.B.E., R.A., R.W.S.

Designed by Andrew Ivett
Typeset in Bembo
Printed and bound in Great Britain by Henry Ling Limited at the Dorset Press, Dorchester, 1999.

Contents

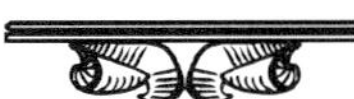

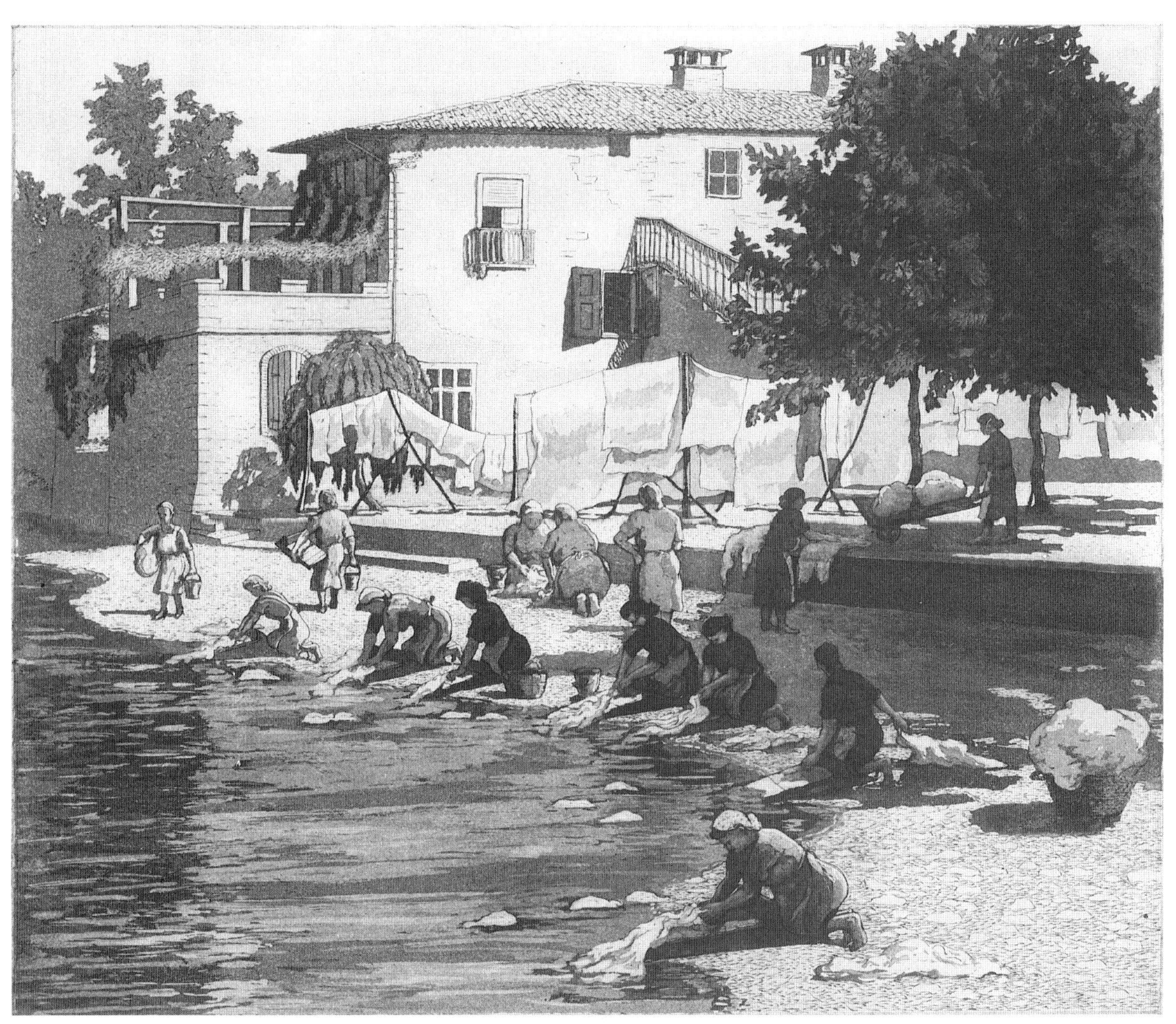

Rachel Ann Le Bas (b.1923)   RE496 *Washerwomen by Lake Garda*   Etching & aquatint 253 x 302 mm

Preface

The Ashmolean Museum's collection of European Old Master prints, which is wide-ranging and has some great strengths, has its roots in the extraordinary accumulation made by Francis Douce and bequeathed to the University in 1834. Some remarkable subsequent additions have come by sporadic gift, such as the Rembrandt etchings from the superlatively discriminating nineteenth-century collector Chambers Hall. It is, however, beyond the Museum's capacity to develop this collection systematically by purchase. By contrast, British prints of the twentieth century, particularly monochrome prints in traditional relief techniques, have been an area in which the Museum has come to specialize and, since the gifts of Arthur Mitchell in the early 1960s, systematically built up the holdings to a point where they constitute one of the most comprehensive reference collections in England. The Ashmolean Print Room is one of the few in the country able to care for and make available to the public world-class collections of prints and drawings, and is proud of the access it offers to a wide range of visitors.

When, therefore, shortly after I came to the Ashmolean in 1990, I was informed that the Diploma Collection of the Royal Society of Painter-Etchers and Engravers (now called the Royal Society of Painter-Printmakers, but still claiming the historic abbreviation "RE"), a prime historical archive of printmaking in Britain over the last 120 years, could no longer be safely stored, let alone made adequately accessible, at Bankside, and was asked whether the Ashmolean would house it as a long-term deposit, the then Director Christopher White and I were pleased to accept the offer. The Diploma Collection consists of a print given by every member of the Society on election and not only contains some very fine prints but also constitutes a coherent document in the history of art and taste. Some artists were elected early in their careers, before they developed the style of work by which they later came to be known, so that the Diploma work can now seem intriguingly uncharacteristic; Graham Sutherland is an example. The Diploma Collection is therefore kept together in chronological order, separate from, and complementary to, the main Museum collection. To leaf through the collection gives a special sense of developments in artistic fashion and achievement over time.

To adequately sort, conserve, mount, and catalogue this large collection was beyond the resources of the Department of Western Art, but thanks to the labours of several volunteers and the kindness of several benefactors, great progress has been made towards bringing the collection into good condition and order. That so much has been achieved, especially in the formidable and expensive task of conservation, has been due to generous benefactions from the Esmée Fairbairn Charitable Trust, the Pilgrim Trust, TI Group plc, Price Waterhouse Charities Committee, Mrs Lowell Winkelman, and Mr Jolyon Drury, as well as several other smaller donations. The Museums and Galleries Commission funded a conservation internship in 1993-4, enabling Richard Hawkes to carry out some of the more pressing conservation work. Among volunteers who have worked to impose order on the collection are Margaret Corsham, Anne Stevens MBE, and above all Clare Tilbury, Lecturer in the History of Art at Oxford Brookes University, whose listing of members and their diploma works published in this volume is the distillation of a great amount of devoted and meticulous work. Publication of this book was made possible by a grant from the Paul Mellon Centre for Studies in British Art, a consistent and valued source of support for the Ashmolean in recent years. I am delighted to have been able to recruit so distinguished a print scholar as Martin Hopkinson, formerly of the Hunterian Art Gallery, Glasgow, as author of the present history and guest curator of the exhibition. At the Ashmolean, the Director Christopher Brown has been support-

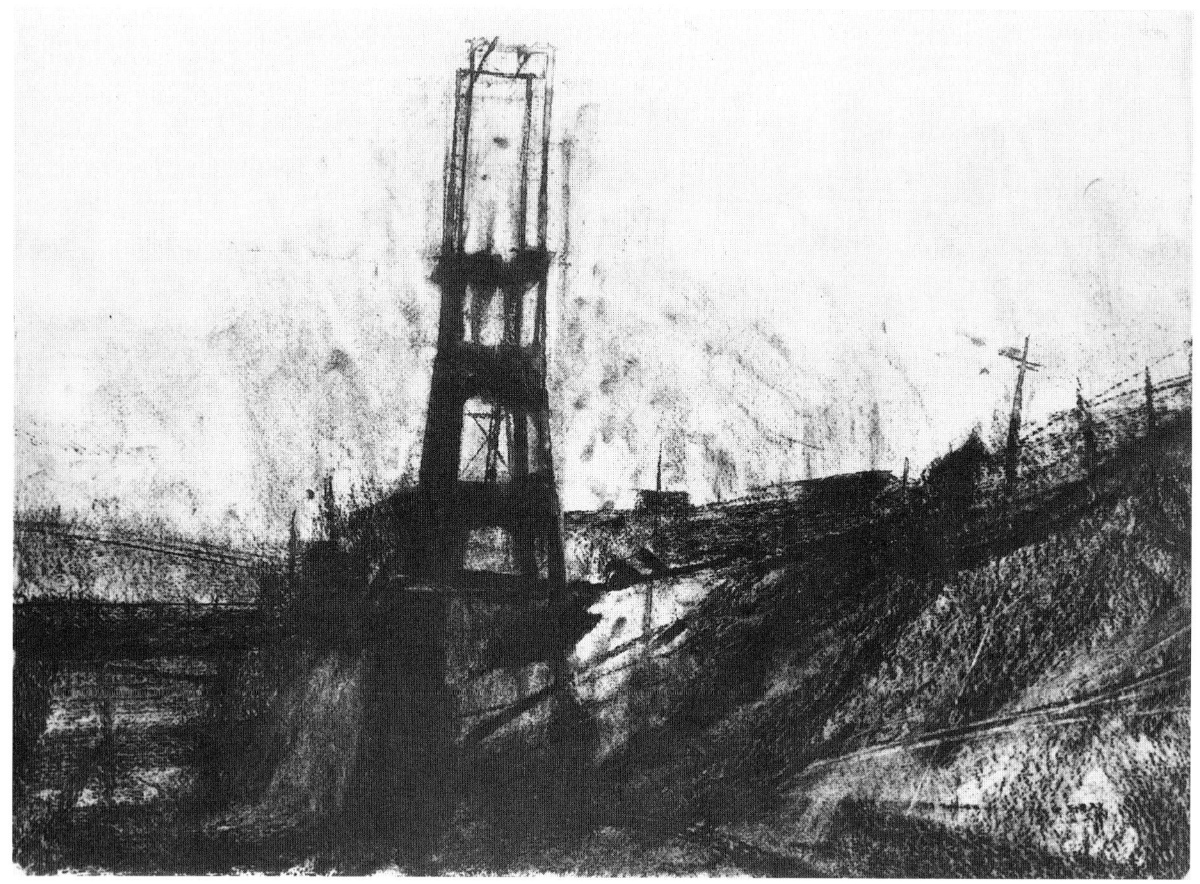

Prof. David Laurence Carpanini, P.R.E., Hon. R.W.S. (b.1946)   RE574 *South Winder* 1978   Etching 350 x 490 mm

ive and Judith Chantry, Colin Harrison, Kate Eustace, Catherine Casley, Ursula Mayr-Harting, Caroline Campbell, and Ian Charlton have all played valiant parts. Ian Lowe, Hon. R.E., former Assistant Keeper in the Museum, who maintains from Cumbria a watchful eye on the Department and on the well-being of the print collection he did much to form, has given constant wise and constructive advice.

Several members of the RE itself have given precious input - notably the present President Professor David Carpanini and the Honorary Curator Vikki Slowe. Simon Fenwick's skilled pro-fessional work on the Society's archive has provided valuable material for research. Above all, Joseph Winkelman, former President, and a loyal friend both to the collection and to the Ashmolean, was instrumental in the original proposal to place the collection in Oxford and has been tireless and self-less in ensuring that the Diploma Collection receives due attention and support. It is to be hoped that the exhibition which is the occasion of this book, in Oxford and on tour, will bring the collection to the wider attention it deserves.

Timothy Wilson, Hon. R.E.
Keeper of Western Art,
Ashmolean Museum.

Introduction

A single man and his campaign to gain official recognition for the work of original printmakers, in particular etchers, dominates the early history of the Society that now bears the title of the Royal Society of Painter-Printmakers. That man was Francis Seymour Haden (1818–1910), a highly successful London surgeon, who was a largely self-educated artist. By his forceful character and his innate artistic talent, he became one of the two most internationally renowned printmakers working in Britain in the late nineteenth century, the other being his great rival and enemy, his American brother-in-law, James Abbott McNeill Whistler.

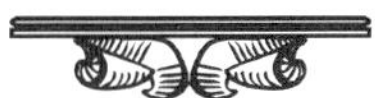

The Status of Etchers before 1880

Before we turn to the foundation in 1880 of the Society of Painter-Etchers, the first name of the Society, it is necessary to outline the causes of the grievances felt by British printmakers over the previous century.

At its foundation in 1768 the Royal Academy limited its membership to painters, sculptors and architects, with the exception of the Italian stipple engraver, Francesco Bartolozzi, somewhat speciously classed as a painter. Although in 1769 as a palliative the Council decided to elect six Associate Engravers, it was not enough to pacify the feelings of British printmakers, who at that time were very largely reproductive engravers. In 1775 a leading figure amongst them, the Scot, Sir Robert Strange, published his vituperative *An Enquiry into the Rise and Establishment of the Royal Academy of Arts* to little effect. In 1809 John Landseer again raised the matter of the exclusion of engravers from full membership of the Royal Academy with its Council, and three years later with thirteen other engravers attempted to involve the Prince Regent in the argument, by sending him a written memorial on the subject. Once more the Academy rejected their claim to recognition on the grounds that 'Engraving is wholly devoid' of 'those intellectual qualities of Invention and Composition, which Painting, Sculpture and Architecture so eminently possess'. For the Academicians, Engraving's greatest achievement was 'in translating with as little loss as possible the beauties of' Painting, Sculpture and Architecture. Throughout the rest of the century they continued to resist the pressure for the admission of printmakers to membership of their body. Although in 1853 the Royal Academy were persuaded by Sir Charles Eastlake to create a new class of Academician Engravers and Associate Engravers for printmakers reproducing the work of painters, there was no recognition of original etchers. The Academy was blind to the growth of interest amongst painters in the late eighteenth and early nineteenth centuries in making etchings and other prints to their own design. This was partly because the most active printmakers were rebels such as Barry, Blake and his followers, or caricaturists, who were regarded as below the salt, or provincial artists like the Norwich School, or Scots, most notably Wilkie and Geddes. Following the example of the watercolourists, who at the start of the nineteenth century formed exhibiting societies to promote their art, etchers began to act in concert. In 1838 the Etching Club was formed and in 1841 it published the first of a series of portfolios. Among its members in the early 1850s were Samuel Palmer, Richard Redgrave, James

Clarke Hook and Charles West Cope. However, most of the products of the Club were derided as pretty and effeminate by the influential critic, Philip Gilbert Hamerton, a long-time resident in France, in his book *Etching and Etchers* in 1868. He felt that many of the etchings were excessively overworked in an attempt to rival painting. Hamerton dedicated his treatise to Francis Seymour Haden. Both men admired the freshness of the prints of the leading naturalist artists in France. They were also impressed by the success of the *Société des Aquafortistes* formed in 1862, and by the advocacy of writers, particularly Philippe Burty, who was quick to recognise the quality of Haden's prints, publishing a catalogue of them in the *Gazette des Beaux Arts* in 1864. Daubigny's *Cahiers d'eaux-fortes* of 1851 and Meryon's *Eaux-fortes sur Paris* of 1852-1854 were important predecessors for the sets of etchings by Haden and his brother-in-law, Whistler: Whistler's *Douze Eaux Fortes d'après Nature,* known as 'the French set', of 1858 and Haden's *Etudes à l'eau-forte* of 1865–1866. Significantly both were published in Paris, as well as in London. The two artists together planned to publish a set of *The Thames from its Source to the Sea,* but their powerful personalities clashed. Whistler resented Haden's overbearing behaviour and was jealous of Burty's praise of prints that the American regarded as inferior to his own. Haden's refusal to meet the costs of the funeral of his medical partner, James Traer, a friend of Whistler, who died in a Parisian brothel, proved to be the final straw for their difficult relationship. The public and financial success of Whistler's 'Thames Set' of etchings, published in 1871, many of which had been intended for the abandoned collaborative publication, had a significant impact in raising the level of public appreciation of original etching in Britain. The series of 'Black and White' exhibitions at the Dudley Gallery, starting in 1876, also heightened the status of the work of the original etcher. A group of younger artists, including Charles Keene and John

Everett Millais, dissatisfied with the Etching Club, had founded the Junior Etching Club in 1857. However, although it attracted talent of the calibre of Whistler to participate in its publications, it had no political impact in furthering the academic status of etchers. Frustrated by the ineffectuality of the Etching Club in altering the Royal Academy's preference for the 'copyist' engraver over the original etcher, Haden resigned his membership in 1878. He also objected to the Club's policy over division of the profits on its publications, which favoured the amount of work done on a plate, rather than strict artistic merit. Hence his sparer style brought less reward than more fussy prints. Haden's rivalry with Whistler may have been the ultimate impetus for the idea of founding a separate society for etchers. It is likely that Haden was riled that the recently formed Fine Art Society had commissioned Whistler to go to Venice in order to make a series of etchings, although it had been Haden, rather than Whistler, who had been singled out by Burty and Hamerton as the outstanding contemporary original etcher in Britain. Whistler's idea for his project may have partly arisen because of Haden's sudden burst of activity as an etcher, after a two-year break from printmaking. For Haden made well over 30 etchings in 1877 and 1878, spurred on by the offer of his first major one-man exhibition of prints by the Fine Art Society, which also commissioned two etchings from him. Furthermore at the end of the same year, he had allowed the Fine Art Society to republish his 1866 *Fine Arts Quarterly Review* article, "About Etching", to accompany their major old master print exhibition, *A Collection by the Great Masters.* The critical attention paid to his Fine Art Society exhibition may also have encouraged him to believe in the viability of a new exhibiting society. It does not seem entirely coincidental that the meeting at Haden's house in Mayfair, 38 Hertford Street, at which the new society was proposed, was held during Whistler's absence in Italy.

Haden and the Setting up of the Society of Painter-Etchers

According to Nazeby Harrington, this meeting took place on 31 July 1880. Present were James Tissot, David Law, William Spread, C.W.Sherborn, Heywood Hardy, Ernest George, W.B.Tegetmeier and George Redford, as well as Haden himself. Of these only Tissot was a significant figure as a printmaker. He held a very equivocal position with respect to Whistler and Haden. Although Tissot claimed to be a friend of Whistler and had probably been inspired by the American to resume etching in 1875 after a break of a decade, he had refused to appear at the Whistler-Ruskin libel trial to testify on Whistler's behalf. Furthermore, like Haden, he was a direct rival of Whistler in the field of etching. Herkomer remembered Tissot's contribution to the discussion as 'to vowing vengeance against the dealers.' Heywood Hardy was a vigorous etcher of animal subjects. The Scotsman, David Law, though prolific, was a mediocre printmaker. C.W.Sherborn worked as an engraver for jewellers and silversmiths, and became a specialist etcher of bookplates. Ernest George was an architect working in the Norman Shaw 'Queen Anne revival' manner, who dabbled in etching. Spread was a minor architectural etcher. George Redford was an art sales correspondent, probably invited along to ensure that the proposed new venture was well publicised. Neither he nor Tegetmeier seem to have been active etchers. Letters of approval for the formation of the society were read from the President of the Royal Academy, Lord Leighton, the Royal Academicians, H.Stacey Marks, Edward Poynter, G.F.Watts and Thomas Woolner (misrecorded by Newbolt as Walker), the sculptor, J.E.Boehm A.R.A., the leading Scottish sea-painter, Colin Hunter, G.P.Jacomb-Hood, Thomas Huson and others. According to Harrington, earlier in July Haden had written to a large number of etchers, and some painters, who had never etched, asking them to attend, and enquiring whether they were in favour of helping to form a society for the promotion of original etching, on somewhat similar lines to the Old Watercolour Society. Many Academicians had been approached and only Millais declined to join, though sending his best wishes. The meeting set up a provisional committee consisting of Haden, Hardy, Hubert von Herkomer A.R.A., Alphonse Legros, then Slade Professor at University College, London, Robert Macbeth and Tissot, with power to add to their number. Legros' lack of English was to be a handicap. According to Herkomer, he could not understand the language, perhaps an exaggeration, and sketched all through this meeting. Haden had alerted the press to the proposed new Society, stressing his prolonged consultation with some of the foremost members of the Royal Academy, as was reported in *The Art Journal*.

At a second meeting in Haden's house on 29 November, Philip Hamerton was elected an Honorary Member of the Society, and Dr Edward Hamilton, a Treasury official and author of *The Engraved Work of Sir Joshua Reynolds*, Honorary Treasurer. Haden proposed, 'that with a view to obtain an adequate representation of the art of original engraving *in all its forms* (Painter Etching) as it exists at present, as well as to provide a constituency out of which to elect the first fellows of the Society, an exhibition as comprehensive as possible of the works of the best living etchers be opened in London at not later than the first Monday in April 1881… ' This resolution was passed, as was another, 'that as the Society is founded for the promotion of original engraving *in all its forms*(sic) etched copies can not be received for exhibition although original works by *copyist* etchers are eligible.' This exclusivity was to remain a central tenet of the Society for 22 years. Of course it meant that those printmakers accepted within the hallowed walls of the Royal Academy were effectively shut out, as their work was by and large reproductive. The term 'Painter Etching' used by Haden echoes the title of the classic multi-volume catalogue of master printmakers published at the beginning of the nineteenth century by Adam Bartsch. The etchers were keen to be seen in the mould of Rembrandt, a great painter as well as the most celebrated of printmakers, rather than to be considered as practitioners of a lesser more craft-like art.

By the date of the next meeting, 14 December 1880, Haden had entered negotiations with S. Weil of the Hanover Gallery to house the show. Six days later the Council accepted Haden's proposal that the Council, two Royal Academicians and one other 'prick off, each on a separate list, the names of those exhibitors who, from the character of their exhibits appear specially to recommend themselves as Fellows of the Society; and afterwards, uniting in Special Committee for the purpose, proceed to select as many of them as may be determined upon of those exhibitors into the Society to constitute the persons so elected, on signing the statutes of the Society, with the provisional Council the roll of the Society and to be *ipso facto,* the first Fellows'. Haden announced on 23 December that, on his own initiative, he had agreed with Mr Weil that the Society would take the Hanover Gallery from 14 March to 31 May. The proprietors of the Gallery would meet all the expenses and allow the Society 15% on all sales. It is not clear what percentage the Hanover Gallery took, but it is likely to have been in the region of 45–50%. Haden also stated, ' that in consideration of a certain opposition to the formation of the Society, he had made a certain communication to *The Times* & *The Athenaeum*, which had been published.' It was also at this meeting that Haden finally accepted the wish of the Council that he should be the Society's first President. He himself later recognised this meeting as the day of the foundation of the Society.

Haden was still anxious not to alienate members of the Royal Academy and the print trade, by appearing to be too exclusive, but also keen to make quite clear his concept of originality in etching and engraving. The Society needed to establish its position. It was not against reproductive artists as such, but planned only to show their non-reproductive work. Around about this time, the Society of Painter-Etchers published a leaflet announcing its first exhibition. As was often the case for a Victorian society setting itself to clarify a situation, the tone was stentorian and the statement prolix. The Society emphasised that ' the Council are fully alive to the excellence of much of the work done by these artists (copyists), and, as etchers themselves, recognise the superior painter-like qualities which, even as a means of reproduction, the etched line has shown itself to possess. On the other hand, they do not lose sight of the fact that the principle of originality which lies at the bottom of legitimate etching determinate its rank and comparative value as a special form of Art. They propose, therefore, without taking cognizance of mere differences of practice (with which they feel that they have nothing to do), to treat *all original work* contributed to the Exhibition strictly on its merits, and, to that merit alone, to open the door to its Fellowship.'

Late in December Herkomer had cold feet about the new Society. He had only relatively recently received recognition from the Royal Academy and had established very good personal relations with the print trade, which he did not wish to endanger. In an undated letter to Haden he worried about how the Society could possibly make ends meet and made the point that ' the copies *always* sell better than any original can'. He was concerned too, both about the reaction of the reproductive engravers in the Royal Academy and the print trade, which was predominantly interested in selling editions of engraved copies after contemporary and Old Master paintings. On 30 December Herkomer raised the question of who was to pay for the framing, the printing and posting of the circular, when the Society started without any money.

'… I am convinced that we shall gain nothing in a pecuniary sense by our society, and I see that we start with unfriendly eyes at us, & if you will allow me to name it, the latter comes from no cause such as you mentioned viz: that some member of the committee has so spread the report that we are going to *turn out* the copyist. Your first mention in public of the proposed society, as Mr Hardy said, would have caused that. I deeply regret the fact of a rival society forming, & cannot believe that the circular will in any way disarm the feeling already started amongst our undoubted rivals. I like the circular now, & only deplore that it did not come out in *that form* quite at the beginning. The trade have so far in no way treated me meanly – & until they do so, I shall deem it my duty to respect them, & can enter no scheme that means defiance.

You know how I would wish to further the cause of original etching, but on consideration I believe the exhibition of works of a society will not encourage new views to talk it up (sic). They *may* just for the one exhibition, but would soon fail to send works; for painting must always remain the principal work of painters.

New men need not search for a place to exhibit in, for London abounds in opportunities of the kind, but they want to know how *to set about* etching & if every etcher made it a *pleasurable* duty to show the first principles of the technique & arouse the enthusiasm for the art in those he teaches, more good would accrue from that, than from any efforts of a society.

I can pledge myself no further than the 'test' exhibition & that is not clear to me in all its details…'

The possible rival society referred to by Herkomer was one of reproductive etchers, founded eventually in 1882, with the confusing title 'The Etchers' Society'. This society, which appears to have been short-lived, was formed by a group of Roman Catholic gentlemen with the intention of publishing prints from time to time, principally portraits (*Portfolio,* 1882, p.84).

In his reply Haden explained that he would himself pay for the postage of the circular, that artists must send their works framed, and that 'we are not, in any way whatever, committed to the proprietor of the Hanover Gallery – or in any sense in his hands.' As for the trade, he argued, 'that the action of the trade & the printsellers' association combined is detrimental to art – the public interest, & the artist, & that the habit which has arisen of advertising the destruction of plates while in a good state is a barbarism, …and a mere trade device…' Haden urged Herkomer, if not satisfied, to withdraw immediately from the Society, before it could be a public and damaging act. Herkomer promptly did so, but in an amicable spirit.

The regulations for the 'test' exhibition were as follows:

'1. All works sent to the Exhibition must be *original*, and if for sale, the *bona fide* property of the Artist.

2. All forms of Engraving on Metal, whether with the burin, the etching needle, by mezzo-tint, or aqua-tint, or by whatever other process the Artist may choose as a mode of original expression, are understood to be included in the term of 'Painter Etching', and, as such are **admissible** to the Exhibition.

3. Although a preference will be given to recent and unpublished works, a limited number of fine and rare etchings executed by living Artists within the last few years, will, if of a nature to furnish valuable illustration of the condition of the Art, be admitted.

Note - The Provisional Council, on the occasion of this test Exhibition, though anxious to impose as few conditions as possible, must yet remind exhibitors that a departure from strict moderation as to size of either frame or mount may subject their works to rejection.'

Herkomer's wife translated the circular into German and Herkomer circulated it amongst his German artist friends. On 1 January Haden sent a bundle of prospectuses to his New York friend and patron, the dealer-collector, Samuel Putnam Avery asking him to distribute copies to *Scribners, Harpers, the New York Herald, the Tribune* and the recently founded *American Art Magazine* ' and generally make it known'.

At the next meeting on 17 January 1881 Sir William Drake F.S.A., who had published the first catalogue of Haden's etchings the previous year, was elected Secretary of the Society. Drake was a senior partner in a prominent firm of solicitors, director of several banks and politically connected to Gladstone. He was an influential figure amongst connoisseurs as chairman of the Burlington Fine Arts Club, for whom in 1877 Haden had arranged an exhibition and written the pioneering catalogue devoted to Rembrandt's etchings. In 1880, with Sir Henry Layard, Drake had founded the Venice and Murano Glass and Mosaic Company. Drake was to be a very active officer until his death in 1890, particularly valuable to the Society for his legal advice. Also elected to the Council was the Royal Academician, James Clarke Hook, a leading figure in the Etching Club. Haden further proposed Frank Holl R.A., Colin Hunter A.R.A., Briton Riviere R.A., Charles Keene, a founder member of the Junior Etching Club in 1857, and William Ewart Lockhart, R.S.A. as members, and Richard Fisher as Curator. Neither Riviere nor Keene took up membership. Keene, being a friend of Whistler, may have felt that it would have been disloyal to him, if he joined. At the next meeting, held in the Hanover Gallery on 15 March, Haden announced that Edward Poynter, John E.Hodgson and Henry Stacey Marks had agreed to act as assessors from the ranks of the Royal Academy and that another Academician, Charles West Cope, who had been a founder member of the Etching Club, had agreed to join the Council. Cope was one of the few members of that club to receive praise from Hamerton. It should be noted that Haden was very anxious that there should be good relations with the Royal Academy, despite the fact that the idea for the formation of the new society had partly arisen out of the dissatisfaction over the Academy's exclusion of original etchers from its ranks.

At the meeting of 2 May in the Hanover Gallery the new Secretary read out the Memorandum of Association for the constitution of the Society, which was agreed to by all the Fellows present.

'1. We the undersigned have mutually agreed to constitute ourselves, together with such other persons as may from time to time be associated with us (as after mentioned), a Society under the name and title of the 'Society of Painter–Etchers'.

2.The Society has for its main object the promotion of the Art of Original Engraving in all its forms, and the furtherance of the material interests of persons practising that special branch of Art.

3.The Management of the affairs of the Society is vested in a Council of twenty Fellows, to whom is delegated the power to make, and from time to time repeal and remake such Rules and Regulations as in their opinion may be best calculated to give practical effect to the before-named object of the Society, and such Rules and Regulations, when they shall have been approved by a majority of Fellows present at a General Meeting of the Society, shall be accepted by the Fellows as binding on them.

4.The Council is elected triennially by the Society at its Annual General Meeting.

5.The first and present Council … shall continue in office until the General Meeting in January 1884.

6.The Council regulates its quorum and course of procedure; its members hold office for three years, and are eligible for re-election, and during their tenure of office, and until their successors are appointed, have power to supply out of the body of Fellows occasional vacancies in their number.

7.A meeting of the Society is to be held annually in the month of January (and oftener if the Council so determine) and the Council will summon such meeting, by not less than fifteen days' notice, by circular, addressed to the registered address of each Fellow, stating the object for which such meeting is summoned.

8. All matters for decision by the Council of the Society to be decided by a simple majority of those present, the President, in case of parity of numbers, having a second casting vote.

9.The Election of Fellows is rested in the Council at a meeting, at which not less than five of its members are present. The Election shall take place immediately after each Public Exhibition (of which there shall be at least one annually), and shall be concluded on the same principles as those adopted in the first selection of Fellows.

The Hanover Gallery Exhibition 1881

The exhibition was to be an experiment 'for the purpose of demonstrating the then state of the art in this country, and of ascertaining whether the material existed for the formation of a Society having for its further promotion', as explained in the introduction to the catalogue of the second exhibition. The intense rivalry between Haden and his brother-in-law led Whistler not only to boycott the show, but also actively to campaign amongst his friends to discourage them from having anything to do with the new Society. He wrote unavailingly to his fellow American, Otto Bacher, whose press he had been using in Venice, 'Don't dream of it my dear Bacher! – The Society of Painter Etchers is already ridiculous and I intend that it shall die the death of the absurd. So just wait a bit – and send nothing to this Seymour Haden game.' Whistler of course could not contemplate Haden, and not himself, as the figure in the public eye at the head of the painter-etching movement in Britain. Bacher in fact so disregarded Whistler's advice that he showed 18 prints at the exhibition, including no fewer than 15 etchings of Venice.

Haden himself unwittingly soon provided Whistler with a welcome opportunity to disparage his brother-in-law in the press. Unfortunately he did not know the names of a number of the young American artists, with whom Whistler had been associated in Venice, who submitted work to the show. So Haden suspected that Whistler was masquerading under the pseudonym, Frank Duveneck. Taking with him Alphonse Legros, another rival and enemy of Whistler, Haden visited the galleries of the Fine Art Society, just across New Bond Street, on March 17, asking to see the etchings that Whistler had done at their commission. Whistler's contract with the Fine Art Society prohibited him from publishing any further etchings of Venice for a year. Haden therefore suspected a cheeky subterfuge by Whistler to deceive his dealers as well as his brother-in-law. Learning of this enquiry, Whistler went over to the Hanover Gallery with his second, Mortimer Menpes that evening, but missed Haden. So the next day he demanded that Marcus Huish of the Fine Art Society issue a written challenge on Haden. Haden was forced to write a letter to Huish, denying any suggestion that he had intended to charge Whistler with breaching his contract, or that he had mistaken Duveneck's etchings for Whistler's. Whistler made the most of this contretemps, publishing the correspondence between Huish and Haden in a pamphlet, *The Piker Papers.* Whistler also asked whether Haden was acting officially and dictatorially as President of the Painter-Etchers, policing prints around the town. Ironically it was one of Whistler's friends, Gertrude Blood, Lady Colin Campbell, who had acted for Duveneck in submitting Duveneck's etchings for exhibition, a fact that may have aroused Haden's suspicions.

On March 29 Duveneck himself wrote to Haden from Florence explaining, 'these are the first etchings I have ever made except *two* or *three* small attempts that I destroyed.' Haden picked out 'the remarkable etchings of Venice by Mr Duveneck & Mr Otto Bacher' in a letter of 27 May to George Redford of *The Times.* He went on to write of 'the absence of the etchings of Mr Whistler which the Council much regret but which does not depend (as universally reported) on his never having received an invitation.'

Despite this unfortunate and much publicised incident, the exhibition was an undoubted success. 491 prints were shown at the first exhibition including many of the best contemporary etchers working in America. Haden had certainly heard of the first major exhibition of contemporary American etchings, organised by Sylvester Rosa Koehler for the Museum of Fine Arts, Boston, in the spring of 1881, as the show included 24 etchings by Haden himself, alongside 549 American prints, 50 modern French works and a group of 80 Rembrandts. There was an enthusiastic response to the copies of the prospectuses circulated by Avery and the work of artists associated with the New York Etching Club was particularly prominent in the Hanover Gallery. Exhibitors included the President, immediate past President and Secretary, Robert Swain Gifford, James D. Smillie and Henry Farrer. Other distinguished exhibitors, whose presence enhanced the status of the Society, were the widow of the President of the Royal Academy, Lady Eastlake, the art historian, and the celebrated French battle painter, J.L.E. Meissonier. Some artists used agents to forward their work. Dowdeswell acted for both Bacher and Storm van 's Gravesande. The artist Edwin Edwards acted for L'hermitte. Edwards' wife was the long-time British agent for Henri Fantin-Latour.

The quality of work by the foreign artists prompted the Council at their 7 May 1881 meeting, held in Drake's house at 12 Princes Gardens, to invite 18 of them to become members, among them, the Americans Bacher, Frederick Stuart Church, Duveneck, Gifford, Thomas and Mary Nimmo Moran, one of the first two women Fellows (the other being Dorothy Tennant, daughter of the M.P. for St Albans and later to be the wife of the explorer Henry Stanley), Stephen Parrish and Smillie, the Frenchmen, Félix Buhot, Paul Renouard and Léon L'hermitte and the Dutchman, Charles Storm van 's Gravesande. A high proportion of the early British Fellows had been pupils of Legros, including William Strang, Harrington Mann, Alfred Hartley, Charles Holroyd, the Countess Feodora Gleichen, Elinor Hallé, Dorothy Tennant, T. Irving Dalgliesh, G.P. Jacomb-Hood, G.W. Rhead, Herbert Dicksee, George Gascoyne, Vereker Hamilton, Lilian Hamilton, Gerald Robinson, Alfred W. Strutt, Robert Bryden, Ethel K. Martyn and Robert Spence. Herkomer's influence was much less strong, perhaps reflecting his ambivalent attitude to the Society. Frank Sternberg, Ernest Stamp and probably Hinchcliff were his students, whilst C.W. Mansel Lewis, a Welsh landowner and amateur painter, was a close friend and patron of the German-born artist. Many of the younger artists selected were

contributors of prints to the periodical *The Etcher*.

In all 54 more artists were invited to take up membership of the Society. The Secretary wrote to each advising them 'A Diploma Etching by each Fellow must, on his election, be deposited with, and will become the property of the Society', specifying the number of the work in the exhibition that the Council had selected to be their Diploma etching. The diplomas eventually issued to the Fellows thus selected bore the significant motto '*ne desilies imitator*' (Don't stoop to be an imitator).

Haden did not wait until the end of the show before writing to Avery on March 30 in a letter that Avery passed onto S.R. Koehler, editor of the new *American Art Review*. Haden asked Avery to convey to the American etchers '...our sense of the excellence and number of their contributions to the exhibition… We have been fairly taken aback by the quality of these works, no less than by the promptitude and spirit with which our invitation has been responded to by our brothers on your side of the water. If the exhibition does no other than to make known the feeling and talent which undoubtedly animate the American etchers, it will not have been in vain. But it will do more than this. It proves to demonstration(sic) the ample material which exists for such a society as we have contemplated, and to the final organisation of which we shall now immediately apply ourselves. The display is all the more remarkable considering the very short time – scarcely three months – since our invitations were issued. Our gallery of eighty feet long and thirty feet wide is too small to contain more than two thirds of the works sent .We have, however, managed to hang nearly all the Americans, and I am sure our artists will be much struck by them.'

In another letter to Avery, Haden wrote, ' on our side we are quite incapable of this sort of spontaneity of action, and my temperament chafes at it, so that I think after all I shall have to end my days in America.'

One of the most detailed reviews of the show was written by the young Cosmo Monkhouse for *The Academy*, 9 April 1881, p.268. He singled out the

President: 'In pure etching Mr Seymour Haden holds the first place easily among English etchers of original landscape impressions. His set of dry points show what can be done with pure line and burr: they are masterpieces of stalled *impromptu* (sic): with every touch a certain contribution to the realisation of the mental vision. In rapid record of pictorial impressions, with the utmost economy of selected line, the skill of Mr Haden is almost unequalled, and, what is more, his impressions are always worth the record.'

The only other artist that he discussed at length was Legros. 'That the purpose of etching is as suitable for the record of conceptions as of impressions is shown by the very remarkable plates of Mr Legros, whose genius scarcely meets (except among artists) with the recognition it deserves. In all of these slight, and, as it may appear to some hasty and barren designs, there is the germ of a whole picture; a definite suggestion not only of the main line and masses and their relations, but of the scheme of *chiaroscuro* and the quality of the atmosphere. Unless the popularity in England of the pictures of Corot be a mere fashion, these landscape etchings of M. Legros should find a public. There are, however, few persons, however Philistine or wedded to realism, who will fail to be impressed with his design of *Death and the Woodman*, in which every line of wall and tree contributes to the weird surprise of 'sudden death'. For those who are unmoved by the power of this remarkable work there are still left two portraits of the Academicians *Watts* and *Poynter* which they can scarcely fail to admire.' The success of the exhibition probably prompted the formation

in 1881 of a short-lived Society of Scottish Etchers.

The evident interest of American artists in exhibiting with the new Society led to an exchange of correspondence between Haden and R. Swain Gifford, President of the New York Etching Club, as to the desirability of holding joint exhibitions in America and the possibility of a triennial exhibition of the two societies, whilst P.G. Hamerton also wrote to Gifford on 21 November 1881, asking if he had any unpublished proofs of etchings for his periodical *The Portfolio*. Although none of these ideas came to fruition, Haden was undoubtedly encouraged by this American enthusiasm, and in the winter of 1882-1883 he went on an extensive lecture tour of the United States, promoting etching in Milwaukee, Chicago, Detroit and Cincinnati, as well as in the Eastern cities of Boston, Buffalo, Philadelphia, Baltimore and New York. It is likely that the examples of the Hanover Gallery and Boston exhibitions induced the London based dealer, A.W. Thibaudeau, to put together a collection of British prints as part of the large exhibition of the works of French and English painter-etchers for the 13th *Ausstellung* in the Nationalgalerie, Berlin. This exhibition, which included many of the members of the newly formed Society, was reported by *American Art Review* to have been held, 'with a view to giving an impulse to the artists of Germany, who have hitherto neglected the art of etching almost entirely.' That show was extremely successful financially. The Nationalgalerie, Berlin, bought the vast majority of the prints exhibited, in all 302 etchings by artists working in Britain, for £685-7-3.

Exhibitions 1882-1887

As the Society was unable to come to acceptable terms for their second exhibition with Mr Weil of the Hanover Gallery, it turned to the Fine Art Society, which was undeterred by the row the

previous year with Whistler. One new regulation governing exhibits was introduced: ' No works being the private property of a collector, or forming part of any Printseller's stock, or which have been previously exhibited or offered for sale else-where, can, under any circumstances, be admitted.' The Society over the years was to be regularly concerned that works in their exhibitions had not previously been shown elsewhere particularly by print publishers and printsellers. It was suggested, 'that the Fine Art Society should interest themselves in the sale of Etchings exhibited, after the Exhibition, by having a portfolio of etchings to be from time to time supplied by the etchers themselves for the purpose of sale, the terms of sale of each such etching to be 25% of the published price.' It was at the Fine Art Society on 26 January that the Society's first Annual General Meeting was held and an annual subscription set of a guinea, payable on 1 January.

Confusingly, because the previous show had been a test exhibition, the 1882 one was called the *First Annual Exhibition.* American etchers were again to the fore. A further group of Fellows were elected at the Council meeting of 21 April 1882, held on the Fine Art Society's premises. These included two more leading young American artists, Joseph Pennell and Theodore Wendel, another one of the young etchers working in Venice known as 'the Duveneck boys', and the French sculptor Rodin. Rodin's presence was probably due to Legros, who exhibited a portrait of him. William Strang, the Scottish artist was probably the most notable other new exhibitor showing eight works. The young Canadian, Elizabeth Armstrong, later Elizabeth Stanhope Forbes showed two etchings from the address of Mortimer Menpes. The exhibition was less than half the size of that in the Hanover Gallery, 207 prints from 85 artists. Some comments were made in the press on the conspicuous absence of exhibits from some of the members of the Council.

For the 1883 exhibition the Society moved to the Windsor Gallery, 26 Savile Row. *The Portfolio,* 1883, pp.84-85 reported a difference in opinion in the society as to whether artists should be allowed to show previously exhibited prints. It was noted that 42 of the 79 Fellows had not submitted any work, possibly because of this and other disagree-ments. *The Portfolio* also remarked on the fall in number of American exhibitors. This was possibly due to the expense of submitting from such a distance and the greater buoyancy of the art market in the States. Joseph Knight was praised for his 'mezzotints of Welsh scenery, almost too effective and charming, so satin-like in softness, so appealing in contrast of light and shade are they.' Also picked out were Menpes' 'vignettes of characteristic figures, peasants, children – in a group, or more often single; drawn with delicate truth and in the most precise and comprehensive facility of expression'. The most extended appreciation was devoted to William Strang for 'eight characteristic etchings of figure subjects, treated in the style of the early masters, very broadly and simply, with strong emphasis, and an intensity of manner which has more than a reminiscence of M.Legros' severity. In the treatment of the homely types so deliberately chosen one traces the study of German and Flemish masters even to the echo.'

The exhibition had very limited financial success, selling just 49 prints for £50-3-6, prompting a letter from Sir William Drake, who had been one of the main purchasers, to Haden on 17 March 1883:

'I find that the public interest in our Etching Exhibition is so small that it is manifest we cannot in future depend upon any owner of a Gallery undertaking a similar exhibition on similar terms.

It becomes therefore essential in my judgement that the whole subject of the constitution of the society, and its exhibitions, should be carefully reviewed by those members of the Council who take any interest in the matter, and I am bound to say my experience satisfies me that these are very few in number.

We manifestly cannot take even the modest sum of £1 per annum from artists, unless we are prepared to give them something in the shape of an exhibition, and this exhibition unless we can make it far more attractive than it is this year, - can only be carried on at a considerable expense.'

Drake was evidently unused to the common failing of many artists on the council of a society to be active in promoting the common good, but his comments did lead to some alterations in the Society. For later in 1883 it was agreed at the Annual General Meeting to enlarge the Council from 20 to 26 and that Fellows of the Society should be allowed to submit two etchings to the exhibition without any judgement on them by the Council. Haden's proposal that there should be an annual portfolio of etchings selected from the Society's exhibition was unanimously approved. At a subsequent meeting it was decided that the portfolio was to consist of 12 etchings to be known as 'The Selected Work of the Society of Painter-Etchers for the year 18—'. The plates were not to be less than 8 inches or more than 12 inches in any dimension. No edition was to be larger than 327, of which 25 impressions were to be reserved for members of the press and two for the Society itself. Each impression was to be signed and 'Remarques', vignettes in the margins outside the main image, were specifically banned. Any process of engraving was permitted. The portfolio was to be offered to the print trade at a minimum net price of 600 guineas. This amazing price may have destroyed any chance of a partnership with a print publisher. The proceeds were to be divided on the basis five–sixths to the artists one–sixth to the Society. The publisher was to bear all the costs of publication and to employ the Society's printer, Frederick Goulding, another student of Legros, who had recently been appointed to that position. The Fellows of the Society would select the individual plates for the portfolio by collective vote. No Fellow was to have more than two prints included in a single issue. Only plates by Fellows that had not been exhibited previously, or offered for sale elsewhere, or commissioned by a third party would be eligible. The Council should have discretion to publish a second portfolio of 12 plates the same year, if there should be sufficient demand. In 1884 insufficient etchings were received for the proposed portfolio to go ahead.

Later in 1883 the Society accepted an invitation to exhibit from the Austrian Commission for the International Exhibition of the Graphic Arts at Vienna. Haden continued to press the Royal Academy for a change in its attitude to original etchers and engravers. He published a paper that he had read to the Society of Arts, *The Relative claims of etching and engraving to rank as fine arts, and to be represented as such in the Royal Academy of Arts*. There he suggested that the Academy follow the practice of the French Salon in dividing 'the art of engraving' into two classes '*l'eau-forte*' and '*la gravure*', with 'a distinct representation given to each.'

At the Annual General Meeting of 1884 the Society felt well enough established to press the claims of etchers on the Royal Academy. Haden was authorised to write on behalf of the Society to complain at the exclusion of 'original engravers' from membership of the Royal Academy, while "the 'Translator Engraver' is admitted to the Royal Academy's highest honours and advantages." He proposed:

'1. That in respect to original engravings (painter-etchings) henceforth submitted for exhibition in the Royal Academy, such engravings instead of being hung as they are at present with unoriginal work shall have on the walls of the Academy as distinct a place as that which is accorded to the other forms of original art.

2. That the authors of such works, if their merit be acknowledged, shall be admitted to membership in the Academy on an equality in respect with Painters, Sculptors, and other practitioners in special branches of original art.'

Despite the presence of 10 Royal Academicians amongst the Fellows of the Society, this plea went unheard.

The Walker Art Gallery, Liverpool, was the first provincial institution to offer to house an exhibition of work of members of the Society as part of their annual Liverpool autumn exhibition. The financial terms were that the Walker took 15% of any sales, the Society would receive 10% and the artists the remaining 75%. The Society's 262 etchings formed just over 10 per cent of the show. Once again American artists featured prominently. This exhibition saw the first appearance of Frank

Short, who was to become one of the Society's most significant members.

In 1885 the Society moved the seat of their exhibitions yet again, this time to the Dudley Gallery in the Egyptian Hall, Piccadilly. This was an appropriate venue, as the gallery had long been associated with 'black and white' exhibitions and in a way the Society's annual exhibitions were their successors. This exhibition saw the first appearance of an etching by Walter Sickert. Duveneck and Pennell both showed groups of Venice etchings. Also notable were four works by the critic, P.G. Hamerton. Strang, Short and Joseph Knight were strongly represented in the exhibition.

From time to time the Society felt that it needed to spell out some of the fundamentals about etching. So at the 1885 Annual General Meeting the Council stated 'every collector of prints must have observed the confusion arising from the want of a clear definition of the words 'proofs' 'Artist's Proofs' 'Remarque Proofs' 'Prints' 'States' etc. It appears to the Council desirable that an authoritative definition of these different conditions of a plate, should, by common consent, be determined upon, so that in future it may be generally adopted. They are now in communication with the French Société of Aquafortistes on this subject, and have submitted to them propositions on the point which they hope may meet with their adoption.'

Legros' resignation from the Society was announced at the meeting on 14 May 1885. His compatriot, Tissot, and Herkomer (for the second time) also left the Society that year, possibly finding it difficult to work alongside the dictatorial President. However, all three were so well established as artists that they had plenty of outlets for their prints and did not need the publicity of the Society's exhibitions.

A number of the original promoters of the Society were not active exhibitors. It was decided at the 1885 Annual General Meeting that the Council should have the power to create a new category of Honorary Fellows. The first of these elected on 30 January 1886 were the critic Hamerton, who though he had made some etchings, was really an amateur, and the artists Alma-Tadema, Cope, Hodgson, Holl, Hook, Marks and Poynter, all of whom were closely associated with the Royal Academy. It looks as if this was in part a gesture of conciliation towards that body. However, at the following Annual General Meeting on 27 April 1887 various rules were introduced to make the Society more exclusive.

Firstly, from now on Fellows were to be elected only from Associate Members. Secondly, exhibitions were to be confined to the work of Fellows and Associates, except only in the case of artists whom the Council may, in their discretion, specially invite to contribute. The creation of the category of Associate membership brought the Society into line with the Royal Academy. At the same meeting the subscription was raised to two guineas and the Council was given power to exclude Fellows or Associates twelve months in arrears.

The 1886 exhibition was held at Derby Art Gallery. The Corporation's invitation had a particular appeal for Haden, as his family had originated from that town, in acknowledgement of which he made an etched copy of Joseph Wright of Derby's 1778 oil portrait of Thomas Haden, his grandfather. Haden himself showed 75 prints, a little retrospective .The Australian, Menpes, was well represented and there was a fine small group of etchings by Sickert, like Menpes, a pupil of Whistler. Sales were extremely poor, only four prints being sold for £11-9-6 ensuring that the Society did not venture to a provincial venue again for many years.

In 1887 the Society returned to London to 160 New Bond Street, above the new galleries of Messrs. Dowdeswell's, a room described by *The Portfolio*, 1889, p.247 as 'dark, & therefore unsuitable for the exhibition of work for the most part requiring close examination. The hangers have, however, done the best they could by judicious arrangement, and have shown on tables and on side shutters on the window recesses delicate or intricate plates.' Particularly noted were Fred Slocombe's 'incisive and effective deliberation of studies', the 'suggestive little plates of J. Pennell & W. Sickert', and the 'silvery finesse of Miss E.A. Armstrong'. Strang received the greatest acclamation 'in his remarkable set of portrait heads, which, for clinching intention, vigorous line and command of the white space, seem to us amongst the most manly of the day.'

An innovation in the catalogue was the indication as to where a print was published by a dealer, a gesture of reconciliation to the trade. Among these dealer publishers were Keppel, Wunderlich and Knoedler, based in New York, Frost & Reed of Bristol, and the London firms, Dowdeswells, McLean, Arthur Tooth and The Fine Art Society.

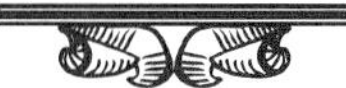

The 'Royal Society'

On 5 June 1888 Haden was able to report that Queen Victoria had made a gracious communication as to the conferring of the title 'Royal' on the Society, an acknowledgement of the status of the Society. The necessary formalities were gone through and the title became official in October. Haden's brother-in-law, Whistler, the President of the Society of British Artists had pipped him at the post, by obtaining the royal charter to rename that society the Royal Society of British Artists the previous year. Haden had been angling for this elevation in status for some time. The election of the sculptress Countess Feodora Gleichen as a Fellow in 1884 may well have been made with a view to her influence at Court, as well as in acknowledgement of her undoubted artistic abilities. Certainly Nazeby Harrington thought that the Fellowship of the Countess and of her fellow sculptor–etcher, Princess Louise, contributed towards the honour bestowed on the Society. Haden had attempted to counter any view that there might be antagonism between the Society and the Royal Academy by writing to the Academician, Hook, offering him the Presidency in place of himself. 'The honest truth is my service to science have been greater than my services to Art, which are confined to the revival (by means of the Society which I have founded) of Painter Engraving; and to the revival (of which I have ample evidence) of the Old Master Exhibitions in the R Academy – in a word to bringing before the public a better acquaintance with the higher forms of Art.' Hook refused on the grounds of his distance from town and the fact that he needed to devote all his time to painting. The granting of the title was an encouraging boost at a time when some printmakers were struggling. The fine mezzotint engraver, Joseph Knight, for instance, had written to resign his membership, saying, 'the fact is there is no demand for any work in mezzotint and I have practically given it up. So it is useless my being a member of the society.'

There was no exhibition in 1888, but early in 1889 negotiations with Alfred D. Fripp, the Secretary of the Royal Society of Painters in Watercolours, bore fruit. From 1889 up to today the Society has shared the same exhibition space as its fellow society starting at 5a Pall Mall East, where they were to remain until 1938. The Council decided 'in partial acknowledgement of the invaluable services of the President (now in his 70s) in promoting the art of the original engraver and in prosecuting the claims to Royal recognition', that 'a strong representation of his etched work be made a prominent feature' of the 1889 exhibition, and at the end of the year proposed that Her Majesty should be written to praying that Haden should be knighted, just as the Presidents of the two Water Colour Societies had been. John Gilbert, President of the Royal Watercolour Society had been knighted in 1872, the year after his election, whilst James Linton, the recently elected President of the rival Royal Institute of Painters in Watercolours had been knighted in 1884. Haden had to wait for another five years for this honour.

The 1889 exhibition was particularly notable for the display of no less than 144 of the President's plates. According to *The Pall Mall Gazette*, 6 April 1889, 'nothing less than a revelation to the many who have seen none but his isolated plates …(they) show how frankly impressionist he is, and yet how completely he is master of line, tone and gradation.' Strang's illustrations to *Death and the Ploughman* were singled out. 'Though reproducing wonderfully the manner and forms of Albrecht Dürer, (they) are more than a *tour de force*, for they are infused with not a little of the true spirit.' Strang's contribution was also reviewed at length in *The Portfolio*, 1889, pp.79-80 , 'We have frequently had reason to mark our appreciation of the etcher's original thought and sturdy power. His faults as an artist are the exaggerations of a sombre imagination, which is apt to dwell on the tragic and pathetic side of life with a painful and sometimes grotesque intensity, and will not avail itself of the sweeter types and gentler graces that are no less legitimate and true means of expression for deep truths. In devotion to the great master of his art, he seems sometimes to forget that Rembrandt's designs were not great by reason of homely or sordid types and actions, but in spite of them. Nevertheless our modern etcher achieves full and manly expression of the dignity and nobility of the labour-worn sons of the soil; and between the earnestness of his thought and the broad directness of his artistic manner there is a fine fitness, combined in allegorical subjects with a grim fancy that is refreshingly *naïve*.'

Sickert exhibited 39 plates, 31 of them grouped together, described by *The Pall Mall Gazette* as 'absurdly out of place in this exhibition, so extremely slight are they.' These Whistlerian vignettes were at odds with the general ethos of the Society. It is conceivable that Sickert was hoping to become a power in the Society, having recently with a group of artists of similar views, known as the 'London Impressionists', wrested control of the New English Art Club from the Newlyn School, the Glasgow Boys and other admirers of the French naturalist painter, Bastien–Lepage. If so, he was clearly thwarted, because he resigned in 1892 and he did not exhibit again until the 1920s. Indeed it was one of the Glasgow Boys who was prominent in the 1890 exhibition, D.Y. Cameron, who exhibited his Clyde set, whilst also to the fore were two other printmakers of a very different stamp from Sickert, Strang and Frank Short. This exhibition was also notable for the selection of over 100 fine impressions of Rembrandt etchings from Haden's own collection. This was the first example of a new policy adopted by the Society, the exhibition alongside the work of the members of prints by an old master.

1890 saw the appearance in print of Haden's first Presidential Address to the Royal Society, published by the up and coming print dealers, Deprez and Gutekunst. Once more he outlined the lack of encouragement to original printmakers from the Royal Academy. He also complained of 'the readiness of the modern Printseller to see in an etching a commercial opportunity for turning it into something else, and, by the publication of plates which have no pretension to be called etchings, to apply the etching process to purposes for which it is altogether unfitted & with which, in fact, it has nothing in common.' The solution he saw was, ' for the Society to publish its own works, or to arrange with some publisher or publishers, to exhibit, advertise and deal with them.' For 'the publisher's interest is to become the purchaser of the plate, and after taking a limited number of impressions from it, to destroy it, and by this (sic) enhancing the rarity of the impression sell it to the public at a much larger price than would satisfy the artist. Obviously, this is good neither for the artist, nor for his art, nor for the public.' He advocated the artist acting through the Society. 'The first impressions ... the public expect to see *here*, and come here to buy them. The exhibition closed – and not till then – he sells as many *impressions* to the Printseller at such a fair price as will enable any honest tradesman to continue their sale and make a good profit out of them – or place them with him for sale on a proper percentage – but his *plates* and the *copyright* which attaches to his plates, he keeps.'

For the 1891 exhibition all the plates of Turner's *Liber Studiorum* were borrowed, as well as other examples of his engraved work, 'as exemplifying the important influence of the Painter on the engraved production.' Of the members the *Manchester Guardian* praised 'a number of delicate and beautiful river-side subjects' by Frank Short, including 'an unpublished *Entrance to the Mersey*, revealing like most of his production – but not to an excessive degree - the influence of Mr Whistler.' The critic of *Star*, 24 March 1891, thought Frank Short to be 'HEAD AND SHOULDERS ABOVE every other Englishman', and admired the work of the Dutchman, Storm van 's Gravesande. 'His *Lagunes near Venice, Flushing* and *Fishing Boats,* are three excellent studies in different effects of light, given with the least possible work, and yet displaying the most effective results.' His work was also praised in *The Observer*, 17 March 1891, for 'great immediateness of expression and a true feeling for his art, though his drawing of clouds is somewhat monotonous.'

The most consistent attention, however, was paid to the etchings and mezzotints of Strang. The *Daily Mail*, 2 March 1891, wrote 'No one shows more talent and more confidence in his own power than Mr William Strang. Through some peculiarity of temperament, the most painful and ghastly subjects seem to take the strongest hold on his imagination, and the force of concentration and realism which he brings to the expression of his thoughts attracts and at the same time repels the spectator. In *The Cause of the Poor,* for instance, he depicts a group of crippled and deformed types of humanity; in another composition, called 'Castaways', we have a horrible tale of the sea. One of a boatful of men has died of starvation and another is shrieking in wild despair. The larger etching *Charon*, a chapter from the *Inferno* is remarkable for the vigorous drawing of the figures; but it is almost a relief to turn to *The Hedger* at the other end of the room, in which Mr Strang has allowed himself a moment of Millet-like calm.'

The critic of *The Architect*, 7 March 1891, felt that 'the work of Mr William Strang is likely to blot out other impressions. At short intervals round the walls, amid the picturesque of common day in architecture and landscape, the fantastic dream of this artist intrudes itself and arrests the eye whether to vex or delight. When a consummate master renders the material of common day, the art of rendering can hold its own with the art of imagination; but there is nothing here of that order that can quite challenge these works with the stamp of creation, creation of the troubled night.'

On 14 May on the motion of the amateur etcher and major collector of prints and drawings, J.P. Heseltine, the Society agreed that the Fellows and Associates be asked to contribute an etching apiece to be presented to the President as a recognition of his efforts on behalf of the Society. At the meeting of 19 June it was decided that Robert Dunthorne should be appointed the Society's Official Publisher. On 23 July the further decision was made that his commission on sales should be raised from 15% to 25%. Dunthorne was to continue in his position until 1898 when the Society decided that the experiment of being tied to a single member of the print trade was not a success. The decision to appoint Dunthorne, the dealer who already represented a number of etchers who were members of the Society, stemmed from concern about poor sales and the Society's lack of marketing expertise. The employment of a professional had an immediate result in the doubling of sales.

The 1892 exhibition included a display of etchings and preparatory drawings for them by Van Dyck lent by Heseltine, Richard Fisher, Morrison and the dealers Deprez and Gutekunst. The show was notable for a much higher proportion of drypoints, perhaps submitted in the knowledge of the presence of Van Dyck's portraits. Introduced by G.P. Jacomb-Hood, the French etcher, Paul Helleu, exhibited for the first time. The *Daily Telegraph*, 8 March 1892, highlighted Herkomer's portrait of the President, 'a talented presentment, with the face in profile and having all the tenderness and feeling of a highly–finished drawing.'

In the summer Herkomer's lectures as Slade Professor at Oxford University were published as *Etching and Mezzotint Engraving,* giving a further boost to the painter-etcher movement. Of the Society he wrote, 'Though not a member, I have watched its progress with much interest, and distinctly think that it is right and wise that such a society should be formed for the painter etcher's art alone. It has now, I believe, a much firmer basis than it ever had before … they have at last got over their trade difficulties by the appointment of a master printer (Goulding) and also a publisher (Dunthorne), with both of whom they have agreements which, while fairly protective of their interests, are liberal to the trade and 'clear of the abominations attaching to the Printsellers' Association.' Herkomer saw as ' the most serious' difficulty facing the Society, 'for the most immediate present the dearth of good original etchers, from whom they can add to their ranks.' Inevitably this is a problem that has faced the Society throughout its history, there always being a limit as to the number of outstanding printmakers active in Britain during a single period.

In the 1893 show 20 etchings by the young Scottish etcher, D.Y.Cameron, attracted the attention of *The Portfolio*, p.viii, 'in spite of its evidence of intelligent study, his work has plenty of originality. Still more has it vivacity, coherence, and satisfaction with what the etched line can do best.'

The 1894 exhibition was marked by the reappearance of Legros, whose 17 plates stole the show. For the *Manchester Guardian*, 20 March 1894, 'the group of them tells with extraordinary distinction, like fine poems, planted amid the gossip of a society paper. Professor Legros' design has the amplitude of the great age, and its parts are now compacted with the energy and heat of a real imagination. His science of drawing is of the great French tradition, the masters that inspire his composition are the noblest, and the images he delineates are the grandeur of gloomy and threatening scenes, miserable mankind and triumphing death. A mind so sternly attuned and an eye and hand endowed for such noble expression intrudes strikingly among the pretty and fussy and dull plates that make up the bulk of the exhibition.'

The distinguished Scottish critic, R.A.M. Stevenson, devoted an extended review to the French master in *The Pall Mall Gazette*, 15 March 1894, 'In these examples of M. Legros' art the imagination is at once virile and extraordinary. The intensity and decorative majesty of the ideas are supported by the most grandly direct means and by a classic science of composition and general effect. *Le Triomphe de la Mort – Après le Combat* (90) is indeed also a triumph of the imagination. The space is filled with great lines swinging in a decorative rhythm, but everything seems instinct with meaning and sentiment. What energy, what conviction of purpose, what belief in the power of simplicity this etching shows! It is no mere exercise in style. Look at Death reaching forwards in his fatal charge, how every limb, from the hand grasping the banner to the horse's hoofs, carries the conviction of irresistible force. Natural beauty and a suggestion of mystery soften the grandeur of such imposing landscapes as *La Ferme de Brieux, Effet d'orage* (87), *Paysage de Bateau* (103) and many more. These arrangements of landscape recall the dignified art of Poussin, while they are simpler, less artificial in composition, and more alive with suggestions of mystery in the woods, menace in the cloudy depths of the sky, and enveloping light and air throughout. Mr Legros binds the whole work together as Beethoven did in music till every part seems instinct with meaning, and details and lines and tones are exalted by their connection with a noble central purpose.'

The *Observer*, 1 March 1894, noted 'two distinct schools of etching, the school of detailed statement, of which Mr Axel Haig is the greatest exponent, and that of the suggestion, the school of Whistler in his later manner and of Dr Seymour Haden, when he was a working etcher.' G.P. Jacomb-Hood published an article after the end of the exhibition in *The Studio,* 15 June 1895, praising the drypoints of his friend, Helleu, in the show for 'the verve and energy of the original conception of the artist's mind, drawn with a delicacy and sureness of eye and hand akin to that of a Japanese draughtsman.' He went on to note 'a fastidious seeking of the unconventionally beautiful and an expression of it, in a manner that does not smack of the schools, though it shows a hard and severe training of the eye and hand, and no sparing of strenuous study.'

Debates over Widening the Scope of the Society

At long last, in 1894 the President received a knighthood, an award which he felt overdue, complaining to his friend, J.C. Robinson (himself knighted in 1887, Surveyor of the Queen's Pictures and an etcher), of the earlier Royal recognition given to Presidents of 'mere offsets of societies like that of the Institute of painters in water colours!' (J.C. Robinson papers, Ashmolean Museum). Haden's suggestion at the Council meeting of 4 May 1894 that the title of the Society should be altered, might be connected with this honour. His supporters at court may have encouraged him to try and broaden the base of the Society. He argued that etching was only one of a number of methods, which were all covered by the French term *gravure*. In the ensuing discussion C.O. Murray asked whether lithography would then be admitted. Though Haden's proposal was supported by Short, there was a clear division of opinion, Strang and C.J. Watson amongst others feeling strongly that the title 'etchers' should be retained.

At the meeting of 28 February the following year there was discussion about whether the Society could accept for exhibition the prints made by Herkomer in a new technique that he had devised, a form of photo-engraving taken from monotypes, which he called the Spongotype. Although two prints were exhibited, a disclaimer was added in the catalogue, which aroused the ire of the artist. The Society's wariness over accepting Herkomer's new technique may have arisen partly due to the recent memory of another controversy over Herkomer's 'prints'. Whilst Slade Professor at Oxford University, he had published a book illustrated with pen and ink drawings reproduced in photogravure. These he had allowed to be described and sold as 'proof etchings.' Joseph Pennell and others had promptly attacked the German artist for misrepresentation in the press.

The exhibition received a series of very critical reviews, particularly regretting the absence of certain leading printmakers such as Whistler, Théodore Roussel and C.H. Shannon. The *Star*, 26 February 1895, for instance, wrote, 'But despite Haden's best endeavours, the society scarcely comprises all the eminent etchers. The truth is one is astonished at the absence of many names. Where is Mr Whistler's for example? This may be an institution with royalty for a patron; it is a pity it has not more artists for members…He (Haden) created the society. He has compelled many people to become members of it. But he also drove out all the obnoxious characters who once belonged to it, obnoxious, that is, from his point of view. In fact he is the society.' A letter from 'A Painter-Etcher' to the Editor of *The Times*, 9 March 1895, defended the Society by an explanation of its aims which, 'have been from the first, simple enough. To replace quantity by quality; to break up the monotony of mechanical engraving and the pernicious monopoly of the mechanical engraver; and, by the creation of new and individual types, to make the engraving of the future an art and the engraver an artist, represent, and may almost be said to sum up, the object of the society.'

The critical reception of the show and the problem of the Spongotype raised further debate in the Council at their meeting of 15 November, as to whether more techniques could be admitted. Nevertheless it was decided that 'metal' should remain in the definition of Painter–Etching, thus to exclude woodcuts, wood engravings and lithographs. In fact lithographs were not to be admitted for over 90 years! The President and Council became more entrenched against innovation and strengthened their control over the Society. Henceforth the President was declared *ex officio* a member of all committees and sub-committees and the Council or its Committee of selection was given the power of veto for the exhibition. The number of Associates was to be limited to 150 and they were to have no part in the government of the Society. It was laid down that candidates for election should submit no less than three and no more than six unframed specimens of their work for examination.

The following year, at the 4 December meeting concern about boosting sales at the exhibition led to the resolution to return for the 1898 exhibition to the original rule which rendered inadmissible works which had been previously published or exhibited. New competition had arisen for the Society with the foundation of the International Society of Sculptors, Painters and Gravers, with Whistler at its head. The drop in income was probably responsible for Haden suggesting that 'it might be well to supplement our annual exhibitions by such an addition of our reproductive work as would show what an original training had done for us' at a time

' not far distant.' This proposal also reflected the realisation that some of the Society's leading members, including Frank Short, were producing high quality mezzotints after paintings.

From 1890 onwards, sales books survive that give a good picture of the fluctuating financial success of the exhibitions and of the individual artists. Sales, which had reached as high as £455-8-6 in 1890 and £610-15-6 in 1892, dipped dramatically in the mid-decade. The principal patrons of the exhibitions during this period included J.P. Heseltine, H.S. Theobald, and the President himself, the last two of whom had been acquiring prints from the exhibitions from at least as early as 1883.

1894 saw the first purchase by the Scottish collector Sir John Stirling Maxwell Bart., son of the pioneer historian of Spanish art, whose daughter was to give Pollok House and its contents to the City of Glasgow in 1966. Sir John supported the Society's exhibitions for well over 50 years. Other notable purchasers included Charles E. Lees of Oldham, whose watercolour collection is now the highlight of the Gallery in his local town, James J. Burra of Ashford, an uncle of the well-known painter, Edward Burra, and the politicians, Sir Charles Dilke and Austen Chamberlain, as well as the Duchess of Sutherland, Lord de Vesci and Lord Hillingdon. Of the etchers themselves, the more well-to-do Colonel Goff and Axel Haig, naturally supported their colleagues. Museum officials started to show an interest at the end of the decade. Campbell Dodgson purchased Strang's *Adoration of the Kings* in 1898 and Sir Sidney Colvin two Watersons in 1901 for their private collections.

The most successful artists in terms of sales at the beginning of the decade were Herkomer, Short, Strang, Axel Haig, Colonel Goff and R.W. Macbeth. Helleu's *Belle Epoque* prints were extremely popular with purchasers once he started exhibiting with the Society. Herbert Dicksee was much sought after in the second half of the 1890s.

The issue of reproductive mezzotints was raised again in February 1898 as the Society was faced with competition from the newly created Mezzo-tint Society. From the 1880s onwards there had been an astonishing rise in the price of historical mezzotint portraits, encouraged by the publications of John Chaloner Smith, the 1881 Burlington Fine Arts Club exhibition of 198 mezzotints, illustrating the history of the medium up to 1820 and the avid collecting of men like William Meriton Eaton, later second Baron Cheylesmore.

In the same period there was increasing interest in the art market in the original Georgian and Regency oils, which the mezzotints reproduced. Publishers began again to commission contemporary copies in mezzotints of historic portraits. Several members of the Society were now active in this reproductive printmaking. So it was resolved that ' with a view to the gradual introduction of reproductive engraving as a legitimate addition to the work of this society, a limited number of mezzotints, both original and non-original, executed by its members at any time since its foundation, be made a separate feature of the forthcoming exhibition.'

The challenge of the increasing interest in reproductive printmaking, an interest stimulated by the Printsellers' Association which was keen to promote large etched copies of landscapes, was also clearly responsible for the Society's decision to alter its name to the Royal Society of Painter-Etchers and Engravers. This change received Royal approval in time for the 1898 exhibition. Haden was now 80 and his growing frailty was recognised by the creation of a Committee of Council, the three Honorary Officers and two others nominated by Haden, 'to relieve him of unnecessary fatigue.' The influential part played by Frank Short in these changes is indicated by the invitation to him to make a list of mezzotint members of the Society and to deal with the mezzotints in the exhibition with the assistance of Axel Haig. The show eventually included 88 mezzotints, 35 of them by Short, only four of which were not reproductive, and ten by Haden himself. That some members of the Society were unhappy about the precedent set is indicated by Haden's note in the catalogue, that the exhibition of 'finished and unfinished, original and un-original ... is *not a departure from our usual methods.'* It seems probable from subsequent events that the objectors included William Strang, who certainly took exception at the meeting of 10 November 1898 'to the admission of unoriginal engravings to future exhibitions.'

Certainly, not all was harmonious in the Society. After the Annual Meeting of 31 March 1898 Haden drafted an Agenda paper for the next Council including two articles, which were clearly directed against an opposition group. The exact cause for Haden's anger is obscure, due to the fact that the minute books throughout the Society's history very rarely give any detail when there was acrimonious discussion. In this case someone, although not the Council, as Haden wrote in the Minute Book, crossed out the parts printed in bold. The articles in question were as follows

'3. That whereas certain of the earlier Rules have for their **apparent rather than intended** effect to deprive the President of all administrative power; and whereas by virtue of his office, he has for many years successfully employed that power in carrying on, *de die in diem,* the current business of the Society, and whereas his right to do so has for the first time in the course of the last year been called in question **and on more than one occasion openly defied,** the Council do now, and pending the Report on the subject of the Standing Committee on Rules, confirm him in possession of all such disciplinary and administrative power as in the interest of order and good government he has hitherto been accustomed to use **and whereas during a year of *exceptional disorder* he has on one notable occasion had to employ exceptional means to correct it, they do hereby fully approve and confirm him in the use without any exception, of all such means.**

4. To take such preliminary steps as may be necessary to nominate a President Elect.

At the meeting chaired in Haden's absence by his nominee, William Urwick, the Council ignored article 3 altogether and postponed discussion on 4, sending Haden an expression of regret that 'the authority of the Chair was not sufficiently supported' 'on March 31'.

The 1900 Annual Meeting resulted in a fresh debate on the issue of reproductive engraving. Strang, Haig, Charles Holroyd and Ned Swain opposed Haden, Short and Macbeth, but nevertheless the Council's recommendation was carried by eleven to five. Henceforth the Council had power 'to include in any exhibition of the Society some proportion of reproductive work by members, and such examples of old masters' work, either original or reproductive, as they from time to time consider advisable.' The battle was not over as a two-thirds majority was required. The same resolution was passed again by the same insufficient majority at the meeting of 9 January 1902, before it eventually achieved the necessary two-thirds majority at a Special Meeting of 6 November 1902.

A compromise rule introduced on 13 November 1902, 'that every reproductive work submitted for exhibition must be accompanied by not less than three original works' was insufficient to still the opposition to reproductive printmaking. On 12 February 1903 Strang and his friend D.Y. Cameron accepted defeat and resigned from the Society, 'on the grounds of recent changes in the constitution and the scope of the society.' Both were becoming very successful commercially. This fact and the reluctance on the part of some print dealers to allow artists, who primarily sold their works through them, to show with what to them was a rival commercial organisation, was contributory to their remaining aloof from the Society for the rest of their careers. Similar factors may have been the cause for their fellow Scot, Muirhead Bone and his brother-in-law, Francis Dodd, not joining the Society. Strang, Cameron, Bone and two other etchers, Clausen and Augustus John, who never became members, were leading figures in the new exhibiting *Society of Twelve* dedicated to the promotion of prints and drawings that was founded in 1904. Another notable absentee from the Society's lists, probably through loyalty to his master Strang, was the future Director of the National Gallery, Charles Holmes, who had established a reputation as an etcher.

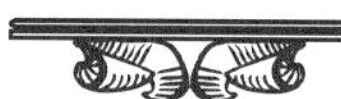

Haden's Last Years as President

The 1899 exhibition included a special display of the bookplate engravings of C.W. Sherborn, a founder member and longstanding friend of the President. Two years later C.J. Watson put together a small display of etchings of the Norwich School, his wife, another member of the Society, the etcher, Minna Bolingbroke, being an expert on the subject. James Reeve, the Curator of the Norwich Castle Museum, was the principal lender.

At the Annual Meeting of 3 April 1902 Haden delivered an extended address to celebrate the Society's coming of age on 23 December 1901. He complained that with the growth over the years in the number of Associates in proportion to the number of Fellows, the Council, the Society's executive, was tending 'to becoming a crystallised and permanent body.' He proposed 5 of the 15 members on the Council who represented London and its immediate neighbourhood should retire every three years and in their place there should be elected five members resident in or near London, who had not been on the Council during the preceding three years.

He also suggested either that a considerable number of Associates should be elected Fellows, or preferably 'a new Rule, providing that Associates shall so participate in the government of the Society as to have the right of voting at the triennial election of Fellows to serve on the Council. By these means new blood would be introduced to the Society and a step would be taken towards doing justice to the Associate members. He also complained about artists not reserving their latest work for the Society's exhibition, but showing with printsellers. There were even Fellows and Members of Council who had exhibitions of their own with dealers during the continuance of, and as if in opposition to, the regular annual exhibition of the Society.

In November a Special General Meeting decided to extend the vote to Associates, not just at elections, but on all questions raised. A further indication that age was forcing Haden to recognise that he should wind down his control of the Society was his appointment of Frank Short and Charles Holroyd to the post, especially created by himself, of Assessor, following his Presidential right to appoint his own officers. The two Assessors were given the right in his absence to occupy the chair in Haden's absence. Haden's action, in effect, marked out for the members his favoured alternatives as successors. For he said, 'I do this in the hope that one or other of them may, at my death, be elected President or Vice-President of the Society, respectively, my experience being that for the proper conduct of its business the two offices are necessary.'

On 4 December Short wrote to Haden, asking him personally to throw his support in pressing the Head of the Board of Education to establish at South Kensington 'a national school of engraving for the furtherance of its art in all its forms: and that a collection of prints be formed in connection therewith (of which the prints in the museum might form a nucleus) for close and continual study by students.' Short felt that his work at the South Kensington Schools, where he had been on the staff from 1891, was limited by the cramped premises that he occupied and that, with Aston Webb's new building now under way, the time was ripe to extend the scope for the teaching of printmaking. Charles Holroyd, as a pupil of Legros and close friend of Strang, represented the opposite wing of the Society. Originally a mining engineer, he was a prolific etcher and also the first Director of the Tate Gallery. Short and Holroyd had entered the membership in the same year, 1885.

Despite his now frequent absence from meetings, the President continued to offer the Society guidance. The 1905 exhibition catalogue contained a note by Haden. 'I have long desired to see Etching treated as more freely and in greater variety as to its methods and possibilities than is our wont – to show how it runs quite naturally not only into varieties of its own but into combination with other forms of the painter engraver's art – especially that of Mezzotint.

I am therefore sending you examples of Etching, *pure et simple*; Dry-point; Etching heightened by Dry-point, *Mezzotint pure et simple*; and Mezzotint enforced by Etching.

You will see that I am sending you nothing but old things, since they are only meant to be looked at for the purpose they are intended to serve; and in the case of the Mezzotints, only as clumsy trial proofs, such as a beginner (which I am), would be making for himself.'

The Society continued to attract some foreign submissions. The fantastic Spanish etcher, Rogelio de Egusquiza, joined the Society in 1898. Eugène Béjot and Edgar Chahine were regular exhibitors from 1899 and 1900 respectively. The Swede, Axel Tallberg, a member from 1891, was joined by Johann Nordhagen from Christiana and Hjalmar Molin from Stockholm early in the new century. Hermann Struck, the leading Berlin portrait etcher, was a very regular exhibitor for many years after he joined the Society in 1902. The Czech etcher, T.F. Simon, became a member in 1910. The Society also admitted promising or already established British printmakers as new members, including Brangwyn in 1903, the Detmold twins, Sydney Lee and Malcolm Osborne in 1905, Martin Hardie in 1907, and E.S. Lumsden in 1909.

There was no shortage of submissions both of candidates wishing to become Associates and of works for exhibition, so much so that at the 1904 Annual General Meeting the Council decided to limit the number of works submitted by each individual member to 9. At the end of that year the French etcher, J.F. Raffaelli, who had been approached to become a member, wrote to ask if the Society would accept two colour etchings for the exhibition, as he was the President of the *Société de la gravure originale en couleurs*. A debate ensued which the traditionalists won, the Secretary writing, 'the Society has never yet accepted engraving or etchings printed in colour.' Alfred East had suggested a compromise that Raffaelli might be willing to send prints in monochrome. 'If, however, you came to submit monochrome prints, a coloured one might be sent as well and the Council could give this matter its full consideration.' Colour prints were to remain excluded until well after the Second World War.

In 1904 the Society's members played a major role in the large exhibition of contemporary English etchings organised by L.W. Gutbier of the leading German print dealers and publishers, Ernest Arnold, in Dresden. In 1906 the Society took up a suggestion of Haden's first made in 1898, and wrote to 12 large provincial towns, sounding them out as to the possibility of housing a duplicate exhibition to their annual London show. An attempt had been made in 1900 to interest the Walker Art Gallery, Liverpool, only to meet with the discouraging response from the Curator, Charles Dyall, 'the Liverpool public did not wish to be educated, it came to the Gallery to be amused! and they wanted colour, and plenty of it!'

Bristol City Art Gallery offered a room to the Society. This out of town show was followed by similar ones in Brighton and Newcastle in 1907 and in Bradford in 1910. These provincial shows were far more successful in terms of promoting knowledge of the Society than in sales. A notably profitable exhibition of 530 prints by 101 Fellows and Associates was also held in Stockholm in 1909 at the invitation of the Swedish Royal Academy of Arts, which had been prompted by Tallberg, an Associate member of the Society. 182 prints were sold, raising £576-13-3 over and above the sales of the Swedish artist members, Tallberg and Molin. This more than matched the sales at the annual exhibitions in London of this period.

Notable collectors in the period up to the Great War purchasing at the Society's annual exhibitions included E.M. Underdown Q.C., Aleco (sic) Ionides, Hugh Hammersley, and George Davison, the photographer . The Society's most regular supporters were J.P. Heseltine and Campbell Dodgson of the British Museum, purchasing for his own private collection. The most successful artists in terms of sales were Helleu, Holroyd, Haig, W.L. Wyllie, Short and Percival Gaskell.

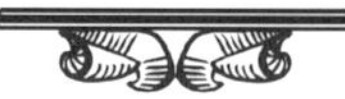

Frank Short's Presidency

Haden died in his Hampshire home on 1 June 1910. Three candidates stood for his post. On 11 June Frank Short was comfortably victorious over Sir Charles Holroyd in the second ballot after Charles J. Watson had dropped out.

Haden bequeathed £300 to the Society. His gold badge of office given by Queen Victoria to him as President was to go to his successor and the chain of office of 72 links of 18 carat gold with three clasps to the Vice President, conditional on Frank Short being appointed Vice President. This last bequest may imply that Haden favoured Holroyd for the Presidency or assumed that the Society would choose a President who had the status of a Museum Director rather than an artist-teacher. Haden's son presented a set of etching tools that had been habitually used by his father. Haden also left £50 to his old friend, C.W.Sherborn, on condition that he introduced into the design for the badge of office 'a fine *pietra-dura* purchased by me at Prince Ponia-towski's sale in Rome 70 years ago with the golden head of her late Majesty executed in pure gold by himself.'

Short soon ran into what was to be a recurring problem for the Society. None of the thirteen candi-dates for the Associateship were deemed worthy of election. The new President had to point out that in his opinion the Council had been judging the work with too high a standard. At Robert Spence's suggestion, the Council then reconsidered the three candidates with the highest votes, with the result that they were elected. Later in the year Legros, who had resigned due to ill health, was elected an Honorary Fellow.

At the 1911 Annual General Meeting, as the Society's finances were so healthy, some considera-tion was given to lowering the rate of commission to 15%, and also to arranging a special exhibition of a notable etcher, and for this artist to deliver a lecture on etching. However, caution prevailed in view of the impending costs of obtaining a Royal Charter of Incorporation for which the Society was apply-ing. Sir Charles Holroyd also pointed out that the Society might want to consider a permanent home. King Edward VII approved the grant of the charter later in the year. In 1912 the Society did decide to lower the commission to 15%.

On 20 October Sir Sidney Colvin, Keeper of the Department of Prints and Drawings in the British Museum, approached the Society for their views on a clause that he wanted included in the new Copy-right Act, then being drafted. He proposed that the Trustees of the British Museum should be given the power to claim one impression of any print published in the United Kingdom and its depend-encies, just as the Bibliothèque Nationale had the right in France. As might be expected Short vigorously and successfully defended his fellow practitioners from this proposal's introduction into the bill.

Further problems occurred over elections in 1911 and early 1912. The Council voted against the election of the distinguished critic, Frederick Wedmore, a Fellow since 1897, and the notable Turner scholar, A.J. Finberg, to Honorary Fellow-ships. They also rejected the candidacy for Associate-ship of James McBey. This latter incident was one of the contributory factors to the decision taken on 25 April 1912 to have a second ballot in elections for Associates when any candidates had received a simple majority in the first ballot. This still did not prevent the rejection of Henry Rushbury, Sidney Tushingham (elected in 1915) and the American, Ernest Roth at the 9 January 1913 election, and Sylvia Gosse (elected in 1917), Roth and his fellow American, Henry Winslow on 8 January 1914. Rushbury had to wait until 1921 for his accept-ance.

Mortimer Menpes, one of the early members, resigned in 1913, after writing what was described as a discourteous letter regarding three works of his that had not been hung and resenting the positions given to his works in past. This is worth noting as in the Society's history there have been remarkably few complaints of this kind. One member, how-ever, the Scot Ernest Lumsden, seems to have had a series of difficulties with the Society's policies, be-ginning in 1913. Unfortunately the minutes of his criticism are so abbreviated, it is not possible to totally understand what were his complaints. In the first instance he may have objected to an amend-ment to the Society's byelaws made at the meeting of 5 March, aimed at excluding old and previously exhibited work. Henceforth, 'in the case of previ-ously published work, only such should be eligible for exhibition as have been exhibited for the first time within 12 months immediately preceding any exhibition held by the Society, to which they are sent.' It is clear from his book on etching that Lumsden also disliked the monotonous displays in some RE exhibitions, which was caused by the twin failures to break up the line of prints into groups and to adopt the practice of hanging from a line at the top. The Society instead followed the norm accepted in contemporary painting exhibitions of using a bottom line to a hang. Lumsden himself preferred the very different and less crowded model of Whistler's print shows.

The Secretary of the Society was relatively ill paid with a salary of just £50 for his efforts on the Society's behalf. On 5 March 1914 the Council resolved that he should be allowed to take a personal 5% commission on all sales.

Over 20 members served their country in the Great War, two of them losing their lives in its course, Wilfred Ball and Luke Taylor, a third, Alfred Bentley, won an M.C., but never fully recovered from his wounds. The Society continued to hold annual exhibitions throughout the conflict, although understandably the average sales each year were nearly halved from the preceding years. Among the new purchasers in 1915 was Queen Mary, who was to be a regular supporter for some years, generally favouring prints of a topographical character. The relief at the end of the war caused sales to leap in 1918 from £268-12-6 to £662-10-0.

In 1915 the Council had made an unavailing approach to probably the three most successful and highly regarded etchers, who were then outside the Society, to Strang and Cameron to resume their membership and to Bone to join for the first time. It seems no coincidence that there was a special exhibition of etchings by Samuel Palmer lent from the collection of Martin Hardie in 1916, the year in which F.L. Griggs was elected to Associate membership.

Short was clearly concerned that the Society was in danger of becoming fossilised. At a Special General Meeting on 20 November 1919 two significant amendments were made to the Society's bye-laws. The artistic strength of the wood engraving revival in Britain was belatedly acknowledged by the decision to end the limitation of acceptable prints to those produced from metal matrices. Secondly a new Bye-law 30A was introduced. 'The Council may notwithstanding Bye-laws 27 and 30 on the proposal of a member of Council, elect as Associate, artists of recognised distinction without requiring the submission of works; provided always that the proposer shall satisfy the Council that the artist so proposed deserves election.'

A second new Bye-law 43 a created a new category of 'retired Fellows' for artists who had not submitted work to the Society's exhibition for four consecutive years. At a further meeting on 28 December the limit of works allowed to be submitted for the annual exhibition was reduced from six to four, a measure intended to increase opportunities for younger less-established members.

Bye-law 30A was pressed into action almost immediately on 8 January 1920 to admit the Society's first wood engraver members, Gwen Raverat and Noel Rooke, as well as the etchers, Hester Frood and Edmund Blampied. Women had been members from the very start of the Society. Nevertheless the decision to admit two women as well as two men with the first use of the new rule may have been an intentional one marking the progress in women's rights in British society as a whole. In the next few years the Council encouraged by Short used the new rule to expand its membership quite frequently. Many of these recruits would probably have been unwilling to submit to the previous election process. The new members included Théodore Roussel and Gerald Brockhurst (1921), Henry Rushbury, John Wheatley, Sickert's pupil, and Claude Shepperson (1921), Laura Knight (1924), E.J. Sullivan (1924) and Sickert (re-elected 1925). An attempt to persuade Muirhead Bone's brother-in-law, Francis Dodd to join in this way failed in 1921-1922.

1921 saw the introduction of the motto *NULLA DIES SINE LINEA* (No day without a line) on the cover of the catalogue of the annual exhibition, something that could hardly have happened in Haden's day, as the phrase had notoriously been adopted by his brother-in-law, Whistler. The changes in the Society may have been hastened by the arrival of competition from the Society of Graphic Art, founded in 1920 by a leading member of the RE, Frank Brangwyn, 'to uphold and maintain the interests of all those forms of art that do not use colour as a form of expression.' This society, which showed drawings as well as prints, exhibited at the Royal Institute Galleries from 1921 to 1940. Brangwyn resigned from the RE in 1920, but does not seem to have induced any other significant defections.

The Print Collectors' Club

The immediate post-war period saw the start in London of the Print Boom, but it failed to extend to the provinces to judge from the meagre return from the Society's 1919 travelling exhibition shown in Bradford, Oldham and Leeds. Nevertheless, the climate was propitious for the formation in 1920 of The Print Collectors' Club, proposed at a Special General Meeting of 29 June. The rules of the Club were approved at another Special General Meeting of 28 October 1920. There was to be an annual subscription of three guineas. The membership, other than Fellows and Associates of the Society, was not to exceed 300 Members, who were to have

(1) the right of free admission to all exhibitions and meetings of the Society

(2) the right of obtaining expert advice on works by submission to one or more members of the Reference Committee

(3) presentation copies of any catalogues, books or pamphlets issued by the Society

(4) to have an annual presentation print.

A selection committee of four, the President, an Honorary Fellow, a Fellow and an Associate was to be elected at each Annual General Meeting to select certain plates for the purpose of making an annual presentation issue, confined solely to the Members of the Club. Not more than 100 impressions were to be taken from each plate, which were then to be destroyed. The first PCC Committee consisted of Short, the Hon. Secretary, the book-plate engraver, J.F. Badeley, the long-time Treasurer, Sir William Plender Bart., and C.M.W. Turner *ex officio,* Martin Hardie, H. Macbeth-Raeburn, William Lee-Hankey, J.R.K. Duff, Frederick Burridge, Malcolm Osborne and Fred Richards. A special bank account was opened at Drummond's. This committee quickly decided by a narrow majority 'that under proper safeguards, those engaged in selling prints, shall in their private capacity, be eligible for membership.' The committee set out the objective of the Club in its constitution as, 'to bring people, who are interested in etching and engraving, into closer connection with those who practise the art, to promote general knowledge of all forms of engraving, and to furnish its members with the opportunity of obtaining expert advice on prints.' It was made out in the Constitution that it was not to be a gentleman's club. It was specifically stated that 'ladies are eligible'.

Although the printmakers themselves as members of the Society were *ex officio* automatically members of the PCC, they were not entitled to receive presentation etchings or other special publications. The Club intended to hold regular *conversazioni* for the delivery of lectures on prints and other artistic purposes. The Committee decided that the Club's first publication would be devoted to Martin Hardie's lecture for the Club at *The Art Workers' Guild* on the British School of Etching. This book by a printmaker, who was also a curator at the Victoria and Albert Museum, was the first in a long series of informative volumes by distinguished museum curators, artists and collectors, which the PCC published, including A.M. Hind of the British

Museum on Rembrandt's etchings, Campbell Dodgson on French etching from Meryon to Lepère, the collector Sir Thomas Barlow on The Woodcuts and Engravings of Albrecht Dürer, Short on British Mezzotints, and Noel Rooke on Woodcutting and Woodengraving. The Print Collectors' Club also decided in 1924 that one proof of each of the annual three presentation plates should be presented to the Print Rooms of the British Museum and the Victoria and Albert Museum respectively. This was a good way of ensuring that the leading printmakers in the Society were represented in the national collections.

The PCC soon came to be a regular welcome source of funds to the Society. Already by 1922 the membership had reached 257, including members living as far afield as Uganda, South Africa, China, Australia and the United States. There were no fewer than 13 members living in Australia in 1923. From 1924, the Club contributed £50 annually towards the Society's rent, a figure increased to £75 in 1926. This was significant, as the Royal Watercolour Society had raised the rent by £75 in 1922, which went up a further £50 in 1926. In addition the PCC from 1922 regularly transferred the balance of its funds to its parent society, the bumper years being 1923 and 1926 when the figures were £661-2-0 and £656-12-9. By 1925 there was a waiting list for membership. The surplus remained between £290 and £475, apart from 1930, a year unusual for the high cost of publishing Sir Francis Newbolt's history of the first 50 years of the society, until there was a dramatic fall in the income in 1937, due to the gathering effect of the recession, to £128-16-3. The following year membership had fallen to 250 and the surplus to £72-15-11 and in 1939 there was no surplus at all, partly due to the increased costs of printing Short's book on Turner's *Liber Studiorum*. The world-wide depression seemed to have less effect on overseas membership which reached 45 in 1938, of whom 26 lived in the United States.

The Inter-War Years

The importance of the Society's annual exhibitions was acknowledged by the purchase of seven prints by the Victoria and Albert Museum. Manchester City Art Gallery bought six prints in 1925 and four more in 1928, the same year that the Harris Museum and Gallery, Preston purchased eight. The leading print dealers, Colnaghi, became regular purchasers through Harold Wright in 1921, who also bought prints for his own private collection. Another dealer to patronise the Society was Mrs Bernard Smith of The XXI Gallery. Well-known collectors to support the Society included St John Hornby, John Charrington, R. Holland Martin, Arthur Acland Allen, Sir William Plender and the curator, Campbell Dodgson, purchasing for his own collection. One of the most active print collectors right into the middle thirties was the banker, H.S. Tuke, surpassed only by the stalwart Sir John Stirling Maxwell and by Campbell Dodgson, who also purchased 17 prints for the Contemporary Art Society between 1932 and 1934, a task taken over by his colleague at the British Museum, A.M. Hind, in 1937. The only museums to purchase in the 1930s were Worthing Art Gallery in 1932, and the National Art Gallery of New South Wales in 1937.

Sales rose from £513-0-6 in 1919 to a height of £1664-19-0 in 1927, only falling under £1,000 in 1931, when there was a dramatic descent to £582-4-6. This prompted a reduction from 25% to 15% in the rate of commission in 1932. Attendances were more constant, ranging from 1605 (889 by tickets and 716 by payment) to 2318 in 1927

(1441 by ticket and 977 by payment). They fell back in the 1930s from 1757 in 1930 to 1103 in 1939. By the end of the decade sales amounted to only £131-16-6 made up of just 30 prints. Over half of the proceeds related to prints by F.L. Griggs, who had died the previous year. From these figures one can see that the collapse of the boom in print purchasing was swifter and steeper amongst the ordinary collectors than amongst the members of the Print Collectors' Club, the devotees of British contemporary printmaking. Towards the end of the 1930s the continuing financial Depression greatly restricted the purchasing of prints. A number of artists found it hard to pay their subscriptions, as is recorded in the Society's minutes.

The 1920s saw a growth of interest amongst British artists in pure line engraving, partly as a result of the exhibitions and publications on early Italian and German masters by the leading print historians and curators of the British Museum Print Room, Campbell Dodgson and Arthur Hind. Robert Austin, later to be President of the Society, one of the outstanding practitioners of engraving, joined the Society in 1921. With the admission of wood engravers and the more open-minded leadership from Short, the Society began to attract large numbers of candidates for membership. There were 45 candidates for associateship in 1924 and 56 in 1925. That year saw the election of one of the finest provincial engravers, Geoffrey Heath Wedgwood, and the first two of the outstanding group of etchers to emerge from Goldsmiths' College, Graham Sutherland and William Larkins, closely followed by Paul Drury in 1926, all perhaps attracted to the Society by F.L. Griggs, a member since 1916, whose work they greatly admired. The Council, however, did not always recognise talent, rejecting Leon Underwood and Edward Bouverie-Hoyton in 1926, and Anthony Gross and Cyril Power in 1927. Of these only Gross was subsequently to become a member, by Byelaw 30A, almost 20 years later.

The Society remained relatively conservative. Artists such as the Nash brothers, Eric Gill and David Jones appear never to have considered attempting to join. All of these had sufficient other outlets to exhibit and sell their work for them to ignore the Society. Very few artists who studied with Leon Underwood or attended the Grosvenor School of Art ever became members either. Electing W. Russell Flint by Byelaw 30A in 1931, however, was hardly likely to encourage the more radical printmakers to attempt to join the Society.

Most British artists, who studied in France in the 1920s or who showed appreciation of Cubism or Surrealism, stayed clear of the Society. Neither S.W. Hayter, nor any of his students or associates at Atelier 17, seem to have had any relation with the Society, until well after the Second World War. This is hardly surprising given that their focus was Paris and the French and European markets. Nevertheless there was no shortage of interest in joining the Society despite the onset of the Depression. There were 34 candidates in 1932. Two leading printmakers left the society in these years, E.S. Lumsden in 1931 and Sickert in 1932. Lumsden had never had an easy relationship with the Society and was now greatly involved, as first President, in the running of the Scottish based Society of Artist Printmakers, founded in 1929. Sickert, who was elderly and not making any prints, simply failed to pay his subscription for four years in a row. If he had had closer relations with members of the Council, no doubt he would have been elected an Honorary Retired Fellow, for which action there were many precedents.

1931 saw a return to an issue recurrent in nineteenth- and twentieth-century print history, the problem of non-original prints. The Society resolved at the meeting of 12 March, that 'no member of the RE should add an autograph signature to any mechanical reproduction of one of his engravings or etchings, or allow any such reproduction to be described as an 'artist's proof.' 1933 saw the election of the distinguished American etcher and engraver, John Taylor Arms, after his somewhat surprising exclusion the previous year when he had submitted works to the Council in the hope of becoming an Associate. Reluctance to admit new Associates amongst his colleagues on the Council may have been the reason for Robert Austin temporarily resigning his seat on 28 January 1936. Neither Bouverie–Hoyton, nor Edgar Holloway had been elected the previous year, not for the first nor the last time.

Throughout the period of Short's Presidency the majority of new members were products of the Royal College of Art. There is no doubt that Short and his successor there, Malcolm Osborne, encouraged the most promising of their students to join. Many of the wood engraver members had been students at the Central School of Art and Design, where Noel Rooke was a very influential teacher. From 1920 there was a strong relationship between the Society and printmakers awarded the scholarship in engraving at the British School at Rome. Leading members of the Society dominated the Faculty of Engraving of the School which chose the successful candidates. REs such as Stanley Anderson, Paul Drury and Wilfred Fairclough were particularly influential in decisions. The vast majority of scholars were elected as Associates of the Society shortly before or just after they were granted the scholarship. Several members too had studied at the Slade. Very few artists starting outside London did not attend a London school, unless they were Scottish. Practically no members had studied in Wales or Ireland. Of provincial schools, Leicester and Burslem seem to have provided the most members between the wars.

Quite a number of members were prominent in the field of art education. Charles Baskett was Principal of Chelmsford School of Art, John Moody, Principal of Hornsey School of Art, Eliab Earthrowl, Professor at Hammersmith School of Art, and Leslie Ward, a long-time member of staff of Bournemouth School of Art. H.P. Huggill combined Curatorship of the Atkinson Art Gallery, Southport with Headmastership of the Victoria School of Art, Southport. Walter Keesey was a HM Inspector of Schools of Art and E.J. Sullivan, an examiner for the Board of Education. Foreign members were rarely admitted, though Hermann Struck, who had been struck off as a member of a hostile country during the 1914-1918 war, was readmitted in 1935, and the Hungarian wood engraver, George Buday, was elected in 1939 and his fellow countryman, Nandor Varga, in 1940.

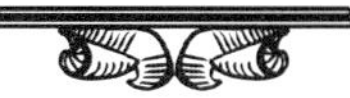

Osborne's Presidency – the Move to Conduit Street

The President, Sir Frank Short, (knighted in 1911), had served for 51 years on the Council and 28 years as President. On 22 April 1938 he announced to the Annual General Meeting that it was time for him to step down, which he would do after reaching his 80th birthday that November. One of the last decisions taken under his Presidency was to agree to move with the RWS to premises at 26/27 Conduit Street, since their home at 5a Pall Mall East was to be demolished in October 1938. The RE was to pay an annual rent of £325. They were initially to have a 21 year lease.

At the Special General Meeting of 2 December there were no less than six candidates for the presidency: Brockhurst, Burridge, William Robins, Rushbury, Hardie and Osborne. Despite the competition, Malcolm Osborne was easily the victor, obtaining a majority over the votes for all the other candidates put together. Hardie, who had served as secretary, promptly resigned from office, being replaced by the wood engraver Iain Macnab, the following July. The new President had been elected an Associate in 1905 and a Fellow in 1909. A pupil of Frank Short at South Kensington, he had succeeded him as head of the etching and engraving school when Short retired in 1924. Of his rivals, Burridge, Robins and Hardie, all were pupils of Short. Brockhurst and Rushbury were both products of Birmingham School of Art. Robins was, like Osborne, an influential teacher, at the Central School of Arts and Design, where Burridge was Principal. Rushbury's work reflected his friendship with two leading etchers outside the Society, Muirhead Bone and Francis Dodd. After the war he was elected Keeper of the Royal Academy, where he did much to reinvigorate the teaching at the Royal Academy Schools. It is likely that Brockhurst's haughty attitude as an artist who moved in high society resulted in his poor showing in the poll.

Concern was voiced at the Annual General Meeting that only 55 of the 120 members had bothered to contribute work to the annual exhibition. No doubt the adverse market conditions for prints had something to do with this. Even the most successful of the Old Master print dealers, Colnaghi's were finding it almost impossible to make any sales in these years. It was agreed for the 1940 show drawings could be included. The Conduit Street Gallery had space on the upper balcony, which enabled them to be shown separately from the prints. Each artist was allowed to submit two new drawings, two drawings that were not less five years old and two working drawings. All these drawings were to be 'studies as may be used for producing prints.' It is evident that colour was shown for the first time in this exhibition, for in 1940 it was stipulated that in future all drawings should be monochrome. The 1940 exhibition went on tour to Cheltenham, Manchester and Sheffield, with very little financial success, no sales at all being made in Sheffield.

Many members were in financial difficulties as a result of the war. In 1942 seven of them applied to pay token subscriptions and in January 1943 the Council decided simply to remove their privileges from the growing number of members unable to pay, rather than to strike them off the membership roll. In 1943 for the first time the Society had to subvent the Print Collectors' Club with £200 to offset the Club's drastic fall in income. The last exhibition during the war in early 1945 was notably more successful than its immediate predecessors raising £470-1-6 in sales and part of the show was toured to the provinces. In December 1945 it was agreed to reinstate all members who had been stuck off the roll before the War for non-payment of the subscriptions on payment of the amount at the start of the war plus a token guinea.

Financially the decade following the end of the war was no more successful than the years of Depression. The Council attempted gradually to broaden the base of the Society. At the 1947 Annual General Meeting it decided to invite members to exhibit books illustrated with engravings on metal and wood, in recognition of the 'prestige of British book production during the last 30 years.' Byelaw 30A was invoked to elect Anthony Gross, Robert Gibbings and John Copley in 1947, Gertrude Hermes and Muirhead Bone (the latter to an Honorary Fellowship) in 1948, and John Buckland Wright in 1953. In 1948 the Chicago Society of Etchers were invited to contribute 50 prints to the Annual Exhibition, an invitation eventually taken up in 1950. For the first time posters advertising the Society were put up on the London Underground in 1949. The Print Collectors' Club had yet to recover its financial position. So in 1950 the Society had to meet the PCC's share of the rent and contribute towards the costs of its presentation volume.

The Society toured its exhibition outside London in 1948, 1949 and 1951, and held a separate exhibition that year in Fulham Library. In 1952 an exhibition was sent on tour to South Africa. Pressure from Associate members led by Charles Bartlett and Harry Eccleston led to a Special General Meeting in 1951 at which it was agreed, 'that Two Associates be co-opted to serve on the Council annually to express their views, but with power to vote, and that one Associate be appointed to serve on the Hanging Committee', as a one year experiment. Bartlett and Eccleston had whipped up Associates to attend the meeting from everywhere, Paul Drury among the Fellows was influential in persuading his colleagues to support the motion. Among public institutions to purchase during this period Aberdeen Art Gallery, Hornsey Public Library, the Art Gallery of South Australia, advised by the print dealer, Harold Wright, of Colnaghi, Canterbury School of Art, Christchurch, New Zealand, the British Council and the Arts Council of Great Britain. The most regular private buyer was the brewer, Arthur Mitchell of Charlton Kings, near Cheltenham, whose fine collection has found a good home in the Print Room of the Ashmolean Museum, Oxford.

1954 saw perhaps the most influential exhibition of non-members' work to be shown by the Society in the course of its history. At the suggestion of the veteran American member, John Taylor Arms, made as early as 1951, the Society of American Graphic Artists was invited to exhibit. Owing to the delay in its realisation, it became a memorial to his promotion of the art of his fellow countrymen. The 169 works included works by many of the outstanding printmakers in the United States. Unfortunately, the currency regulations then in force prevented any of the American prints from being available for purchase. The show included many lithographs and a linocut, media not allowed under the rules for the RE's own exhibitions. Leading recorders of the American scene and way of life such as Martin Lewis, Paul Cadmus, Louis Lozowick, Jolàn Gross-Bettelheim, Benton Spruance, Howard Cook, Samuel Chamberlain and Peggy Bacon, as well as older etchers, like John W. Winkler, and Arms himself, were well represented. However, the most significant presence was probably of a group of artists, who had worked with S.W. Hayter during the time when his Atelier 17 was in exile in New York. Among these were André Racz, Roderick Mead, Karl Schrag, Letterio Calapai, Sue Fuller and Alice Trumbull Mason.

In 1954 there was probably only one other place in Britain open to the general public where you could see American prints of this kind, predominantly abstract, strongly influenced by Surrealism, the little-known collection of James A. McCallum housed in the Print Room of the University of Glasgow. As well as Arms, four other members of the RE exhibited with the American Society, the etchers, Louis Rosenberg and Herman Webster, and the wood engravers, Clare Leighton and Nora Unwin, who had both emigrated to the States.

This show was undoubtedly a major factor in persuading young recently elected artists such as Philip Reeves to expand their techniques and to attract to the Society the South African, Dolf Rieser, an older artist, who had attended Atelier 17 in Paris before the Second World War. Rieser exhibited with the Society for the first time in 1955, as did Michael Rothenstein, both as non-members. For in 1955 and 1956 space was found to admit exhibitors from outside the Society. Rieser was to become an Associate in 1956. The 1955 exhibition was the first at which colour prints were admitted, a decision possibly inspired by the American exhibition. The Colour Woodcut Society founded in 1920 and a secessionist Society of Graver-Printers in Colour founded in 1931, which admitted, in addition to woodcuts, colour intaglio prints, had provided annual exhibitions in which colour printmakers could show between the wars. There had also been eight annual British linocut exhibitions at the Redfern Gallery and the Ward Gallery between 1929 and 1937. However, it was much more difficult after the war for a colour printmaker or indeed an artist working in black and white to find an exhibiting space. The Redfern Gallery was the only dealer regularly to show and sell contemporary prints, until Robert Erskine founded the St George's Gallery in 1954. In fact the success of Rex Nan Kivell at the Redfern and Erskine in marketing colour prints may have been a further impetus to the Council of the RE in their decision to admit colour.

The downside of the American exhibition and the corresponding exhibition of works by members of the RE sent towards the end of 1953 to America for exhibition with the Society of American Graphic Artists, was the financial cost. In March 1954 the Society had to draw on a loan from the Print Collectors' Club to meet cash flow problems. Sales were still very low, just 27 prints were sold at the 1954 show, raising £152-11-0. An additional problem arising out of the exchange of exhibitions was that the RE's prints were misplaced after the end of their American exhibition, which was also shown at the Print Club of Albany and the Library of Congress. It took until 1962 to recover the exhibits, during which time the American Society had closed down. One measure taken in the hope of both increasing revenue and in expanding the scope of the Society, was to hold an experimental open exhibition, charging non-members a submission and hanging fee, in 1955 and 1956. This initiative was due to the persuasive voice of Paul Drury. In 1955 the RWS increased the rent payable by the Society from £325-10-0 to £400.

Also significant for the future was the retirement of Robert Austin in 1955 from the Royal College of Art. Hitherto intaglio printmaking had dominated teaching there and at the Slade. His successor was a lithographer, Edwin LaDell, despite the department still retaining the title 'Engraving School' into the sixties. This appointment naturally reduced the proportion of etchers and engravers emerging from the school. More seriously, a rift occurred due to Austin's unhappiness at the behaviour of the new Principal, Robin Darwin, towards senior members of staff, particularly Gilbert Spencer, brother of Stanley, whom Austin believed had been treated extremely badly. Austin turned down the offer of an Emeritus Professorship on his retirement from the College. These factors combined to end the historic position of the Royal College as the dominant supplier of new members to the Society, a position that it has never recovered. Feelings were so strong that the numbers of promising young printmakers applying for membership fell off sharply and the Society became much more dependent on the supply of students from non-London schools. The department at Brighton where the South African student of Hayter, Jennifer Dickson, elected an Associate in 1959, taught alongside R.T.Cowern, became particularly important in this respect. Also important as a recruiting ground was the West Surrey College of Art at Farnham, where Arthur Hackney taught.

1957 saw a reversal to an exclusive policy for the 75th Annual Exhibition. Sales remained low and it was the turn of the Print Collectors' Club to suffer financially. In November 1958 the Society decided that for the next two years it would meet the Club's annual deficit.

In 1959 non-members were once more allowed to exhibit, including Eric Hebborn, later to be notorious as a forger, whose exhibits of wood engravings after Gainsborough's *The Morning Walk* and *Homage to Breughel* were an indication of his future interests. Twice as many works by Fellows and Associates as by outsiders were hung. 1960 saw a distinct improvement with the sales reaching £677-16-6, the highest figure since 1930, boosted in part by the purchase of 30 prints by the South London Art Gallery, which had agreed to take the whole exhibition as a second venue. The South London Art Gallery was to purchase further groups of prints in 1961, 1964, 1965, and 1967. Another particularly welcome and regular purchaser in the late 1950s and early 1960s was the Stoke-on-Trent Education authority, advised by a long-time RE, Leonard Brammer, who was the Supervisor of Art for that local authority between 1951 and 1969. Greenwich Public Library was also a consistent buyer. Arthur Mitchell remained the single most significant private collector to patronise the Society.

In 1959 Malcolm Osborne had voiced concern at the length of time that an Associate member had to wait before election to Fellowship, largely because of the limit of 50 to the number of Fellows permitted. So at a Special General Meeting on 13 December 1960 it was decided to amend the byelaws creating a new category, Senior Fellow, which would be attained at the age of 75. Nine members had become Senior Fellows as a result by 1962 allowing the election of new Fellows, including a group of wood engravers, James Bostock, George Mackley, Gwenda Morgan and Geoffrey Wales, together with two etchers who were to play prominent roles in the Society, Harry Eccleston and Charles Bartlett. After the South London Art Gallery showing, 88 prints from the exhibition were toured by the Art Exhibitions Bureau to Ilford, Derby, Scarborough, Luton and Southgate. Further measures intended to boost income were taken at the 1961 Annual General Meeting. The commission on sales charged to non-members was increased from 15% to 20% and, to encourage members of the Print Collectors' Club, they were to be allowed a 20% discount on purchases. A welcome windfall came in the bequest of £500 from the estate of the etcher, John M. Aiken.

Robert Austin's Presidency

12 July 1962 saw the resignation of Malcolm Osborne, who was now 82. Each of the three Presidents so far had continued in office into old age. The candidates to be his successor were Short and Osborne's pupil, Robert Austin (then 67), Paul Drury (59), a student of Stanley Anderson at Goldsmiths', and the Birmingham etcher Andrew Freeth (50), who had won a scholarship at the British School at Rome, during Anderson's Chairmanship of the Faculty of Engraving.

Austin was elected on 20 October. The first exhibition of the new Presidency attracted the highest attendance for 30 years and the sales once more topped £600. This reflected a new buoyancy in the print market. Stanley Jones had opened Curwen's lithographic studio in 1958, Kelpra published its first screenprint in 1961 and Editions Alecto was founded in 1962. 'Swinging London' was making its bid to be one of the major centres of the art market. The young 'Pop' artists emerging from the Royal College of Art filled the colour supplements alongside the Beatles.

The culture of youth and adventure, however, was not reflected in the Society. In 1963 Graham Reynolds, curator at the Victoria and Albert Museum, wanted to change the arrangement by which the three Print Collectors' Club presentation prints were given to his museum, as he considered them too staid. He proposed that instead he should be allowed to select one from the whole show as 'the prints chosen by the PCC selection Committee did not fit neatly into the type of print that the Victoria and Albert Museum now collected.' It was pointed out to him that it was the Museum that had been made a member, not the Keeper of the Print Collection, and that the Museum had the right like any other member of the PCC to purchase prints from the annual exhibition at a 20% discount. Other public institutions thought the exhibitions to be of high quality to judge by the purchase of 10 prints by the National Gallery of New Zealand in 1964 and of 22 prints by the Ferens Art Gallery, Hull in 1966. Oxford, Cambridge and the Museum Boymans-van Beuningen, Rotterdam also were purchasers in the 1960s.

In 1965 on the initiative of Anthony Gross, Michael Rothenstein, Julian Trevelyan and other leading printmakers, the Printmakers' Council was formed in London. The aim of this new body was to be more radical than the Royal Society of Painter-Etchers and Engravers. Members were elected for a period of five years only, after which they had to seek re-election. Applicants were to be judged on the strength of their current work. The Council laid out its programme as involving 'opening up markets for members' work; both by organising exhibitions and by collecting and circulating information on exhibitions, collectors and dealers all over the world'. It also circulated data on materials, methods and equipment. The Council wished to be more progressive and non-establishment than the older society and to avoid the air of a gentlemen's club, which was the image of the Royal Society of Painter-Etchers for a number of younger artists.

Initially there was some fear of the new competition. It was probably no coincidence that in 1966 the Society decided to elect S.W. Hayter as an Honorary Fellow, a growing number of his students at Atelier 17 in Paris having joined the Society in the early 1960s. Jennifer Dickson, one of them, a committee member of the Printmakers Council, joined the Council of the Society in 1967. Dickson and artists like Agatha Sorel bridged the two bodies and did much to ensure that there was no conflict between them. Sally McLaren and Dolf Rieser, also Hayter pupils, were co-opted to the Council of the Society in 1967 and 1968, respectively. Rieser became a Council member in 1972 and was joined by Valerie Thornton and Philip Reeves in 1973. Prints were now getting much larger and hence taking up more exhibition space. So on 15 March 1966 it was decided that members, who wished to submit works over 26 x 20 inches in size framed up, could only submit four not six.

1966 was the year of the formation of the Printmakers' Council. Several of the most significant young members of the Society were involved in its foundation. The Council very quickly attracted 160 members. This new body put pressure on for all art schools to set up proper printmaking departments and began creating touring exhibitions. Official backing was signalled by the British Council touring a Printmakers' Council show. There was talk of the Society not allowing artists to be members of both institutions, but fortunately Austin was persuaded by Harry Eccleston not to raise the issue, in view of the danger of not only losing some of the best younger members, but in cutting off the supply of recruits from the graduates of Brighton and Goldsmiths', where Jennifer Dickson and Sally McLaren, respectively, taught. Further opportunities for printmakers were created by the setting-up of the Bradford International Print Biennale in 1968 by Peter Bird, husband of Birgit Skiold, a printmaker who was never elected to the Society.

In 1970 inflation forced the Council of the RE to increase the subscription rates for the first time for 50 years. Henceforth the annual subscription for both Fellows and Associates was fixed at five guineas and the entrance fee for Associates £3.

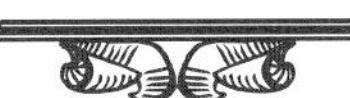

Paul Drury's Reforms

Austin, having reached 75, the age that qualified him to be a Senior Fellow, retired from the Presidency and was replaced at the Annual General Meeting of 26 March 1970 by Paul Drury, the first President for over 60 years to have had no association with the Royal College of Art. He was to be a reforming President, anxious to introduce new blood into the Society. He was also keen to establish closer contact with the members, introducing a regular letter to them from the President, irreverently referred to by his successor, Harry Eccleston as 'the Epistle of Paul.' Until Drury's Presidency, the Society was very much a gentleman's club, very much in the late Victorian mould set by Haden. Malcolm Fry recalls that previous Presidents, following the autocratic tradition of Haden, did not stoop from interfering with the hanging of the annual exhibitions after the hangers had done their job. Drury put an end to this practice. However, as there was still a strong body of traditionalists opposing change within the Society, reforms were gradual.

In 1968 the Council had thought that it would be good for the Society to invite more foreign printmakers to join, provided that they were of a sufficient standard. Two years later the Hungarian artist, N.L.Varga, already an Associate member, was the first to be elected an Honorary Fellow under this new policy. The Genoese woodcut artist, Tranquillo Marangoni, followed suit in 1972. In 1970 the Society received a bequest from the etcher, Joseph Webb, of about £2000, out of which the Joseph Webb Memorial Fund was created. Bursaries, not exceeding £100, were to be awarded annually at the discretion of the Society to a student or students of etching and engraving, who would be selected by the President and a panel appointed by the Council. A first gesture was made that year towards the acceptance of more contemporary techniques. The statement that lithographs and silkscreen prints were inadmissible was amended to read, 'Prints which are exclusively lithographic or silkscreen are inadmissible'. Pure lithography was still regarded by many members as too indirect a method of printmaking to be accepted in its own right.

The Senefelder Club, which had been founded in 1908 as an exhibiting society for lithographers by Joseph Pennell, A.S. Hartrick and F.E. Jackson, held annual exhibitions into the late 1930s, but after the war held only occasional shows. The Society of London Painter-Printmakers, which held its first exhibition in 1948, provided a vehicle for postwar lithographers and artists working in monotype, but it was short-lived. Lithographers had to rely on the Redfern and later the Saint George's Gallery, before the sudden growth of interest amongst art dealers in print publishing in the early 1960s. It was perhaps the association of lithography and screenprint with the commercial art market that was responsible for the cautious attitude of the Society towards lithography, even at this late date. The President at the 29 March 1972 meeting recommended that consideration be given to absorbing the Senefelder Club, but met opposition, on the grounds that the Charter specifically excluded lithography and that the Society had insufficient exhibition space to accept an influx of lithographers.

Concern over the standard of work submitted by the few candidates for election in 1971 led to a proposal by Charles Bartlett, with the support of Joan Hassall, that the age for senior Fellowship should be reduced to 70, to create more space for younger members. Drury, who was now 68, took up this proposal the following year and at the Annual General Meeting it was decided to go further, reducing the age to 60. He was also responsible for abolishing the President's lifetime hold on his office. A motion that 'the President shall serve initially for a term of five years, but may be eligible for re-election after that period' was passed. No less than fifteen Fellows automatically became Senior Fellows as a result of this. Nevertheless only a single candidate achieved the necessary two-thirds of the votes at the next election for Associates in 1973 and disappointment was again expressed at the paucity and lack of quality of the candidates. Even those who had been elected Associates often failed to exhibit in the years following their election, as Jennifer Dickson noted with surprise and dismay at the 1974 Annual General Meeting. In 1972 the Print Collectors' Club decided to abandon their custom of selecting prints from the Annual Exhibition, and instead to return to their original tradition of commissioning prints from members, in the hope of stimulating greater interest among collectors.

1973 saw the renewal for seven years of the Conduit Street lease, but the rent was increased to £700. It was clear that the rise of property values in the West End would make it impossible for the RWS and the RE to remain in the area after 1980. The total assets of the societies did not exceed £3000. So Harry Eccleston was appointed to serve on a select committee set up by the RWS to consider the question of new premises. He was joined by Andrew Freeth, Malcolm Fry, Secretary of both the RWS and the RE, an accountant, Richard Seddon, and Rodney Millard, husband of Jean Harper, a member. Millard was then President of the Advertising Association and had many influential contacts. He was the driving force in the search for new premises.

Consideration was to given to taking a house in Holland Park. However, a chance meeting of the recently elected, President of the RWS, Ernest Greenwood at the private view of his exhibition at the New Metropole Centre, Folkestone, with Sir Gerald Glover of Edgar Investment Limited, led to the offer to include a gallery for the RWS and RE in a new development between Blackfriars Bridge and the Bankside Power Station opposite the Mermaid Theatre. McAlpines had planning permission, but were anxious to avoid the opprobrium in the press that Harry Hyams was receiving over the unoccupied Centre Point tower block. Both the developers and the architects were anxious to include an educational component in their project and accommodation for the two Royal Societies fitted the bill. They offered the Society the space at a much lower rent than the current commercial one.

In 1975 a full economical rent for a gallery the size of Conduit Street in the West End would have been £18-20,000 a year. This figure was likely to rise to at least £30,000 by 1980. There was naturally some concern among the members that a move to the South Bank, at the time a district totally lacking in galleries, from the Bond Street area would have a considerable adverse effect on attendances and sales. However, it was clear that neither society had the resources to afford a space of sufficient size in the West End. It was decided to join the RWS in the new project. £60,000 was needed for a 60 year lease. The gallery rent would be £9500 per annum and there would be a small ground rent. The Council agreed on 3 March 1975 to subscribe at least £1500 towards the new scheme. It was fortunate for the Society that it had received in 1973 a legacy worth £6000 from Dorothea Short, daughter of the second President. Malcolm Fry increased the income by letting out the gallery space outside the times of the two Societies' exhibitions. Originally the RE wanted a 25% share in the new gallery. Eventually having shared the fundraising costs, it was offered a half-share for which it paid about £28,000.

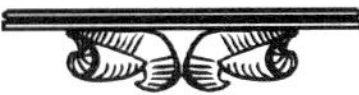

Harry Eccleston and the New Gallery

Drury, having set things in motion for this decisive move, stepped down from the Presidency and at the Annual General Meeting on 26 March 1975, Harry Eccleston was elected to replace him. Eccleston had been a student of Austin's at the Royal College and elected an Associate in 1949. Austin had designed the 'C' series banknotes for the Royal Mint. Eccleston, on the staff of the Bank of England as Artist Designer for their printing works, was to design the watermarks for the 'D' series. Eric Ramsden RE, almost a contemporary, was the chief designer of watermarks for Portals, the papermakers. At 52 Eccleston was much the youngest President in the Society's history to this date. His involvement with the initial search for new premises fitted him extremely well for the forthcoming move and the struggle to raise the necessary funds in an era of high inflation. The subscription rate had to be raised to £10 in 1976 and £20 in 1980.

Fortunately the vigorous campaign to recruit new members for the Print Collectors' Club bore fruit. No less than 50 new members joined in 1975-

1976 and by the Annual General Meeting of 1979, the numbers were up to 350. The discount allowed to members on purchases at the annual exhibitions and a series of demonstrations of different techniques to members of the Club had proved very effective marketing tools. Exhibitions were toured to Harrogate, a venue organised through Eccleston's son-in-law, Martyn Park, in 1977 and the Turner House, Penarth in 1979. The 1970s saw Arthur Hackney follow the example of Brammer in making regular purchases of prints for the Royal Surrey College of Art, Farnham. The most significant institutional buyer was Sheffield City Art Galleries, acquiring 20 prints in 1972, 13 in 1973, 12 in 1974, and nine in 1975. More occasional purchases were made by Hereford College of Art, the National Maritime Museum, the Fitzwilliam and Ashmolean Museums and Portland State University. Private collectors now tended to buy at one or two exhibitions, but not consistently year after year as Sir John Stirling Maxwell, Campbell Dodgson and Arthur Mitchell had done in the past. One Scottish museum curator was an exception to this.

The Society continued to struggle to find candidates of sufficient quality to be elected to Associateship and to retain Associate members as regular exhibitors. 76% of the Fellows and Senior Fellows exhibited works at the 1976 exhibition, as opposed just over 25% of the Associates. The numbers of actual members were falling as a result of the failure to elect new members at a time when subscription income was vital for the health of the Society. Traditionalists were reluctant to allow the introduction of younger more forward-looking printmakers, who might change the image of the Society. It was still necessary for a two thirds majority vote in Council to elect new associates, a rule which Harry Eccleston succeeded in persuading the Society to change to a simple majority vote. Dolf Rieser was probably correct in diagnosing a problem effecting recruitment in the off-putting effect of the old-fashioned image of the Society.

Almost immediately after taking up office in 1975, the new President had unsuccessfully attempted to persuade the Society to consider admitting prints using photographic images and prints enclosed in glass and plastic boxes. On 4 July 1978 following a proposal by Michael Fairclough, a small committee consisting of the President, Charles Bartlett, Michael Fairclough and David Smith was formed out to seek out young, but established printmakers, who would be invited to seek election. The immediate result was a large increase in the number of candidates the following year, when 13, a record number of new members, were elected. Among these recruits were the next two Presidents, Joe Winkelman and David Carpanini. 1979 saw the admission of letterpress into the exhibition and the launch of a Journal of the Society, the initiative of Charles Bartlett, its first editor.

A major set back to the plans for the move was Southwark Borough Council's decision to take control over the site from the private developers. Originally the Societies were to have the basement of the building for offices, educational and storage space. The design of the Gallery interior was very much in the hands of Malcolm Fry, with the Presidents of the two Societies having a very limited input. The slow progress meant that costs rose considerably. At Nicholas Stacey's suggestion, his boss, the management consultant, Francis Singer of Gresham Trust, was approached to help raise funds for the two Societies. The Bankside Trust was formed under his Chairmanship with Stacey, Lord Forte, Malcolm Fry, Harry Eccleston and Ernest Greenwood present at the initial meetings. Later Lord Seebohm and Norman St John Stevas became involved. Delays, familiar to all who have been associated with large-scale building projects in Britain, resulted in the Societies not gaining a foothold in the Gallery until August 1980. There was no annual exhibition that year. Instead a smaller exhibition was held in the Yarrow Gallery, Oundle, thanks to David Carpanini, then on the staff of Oundle School.

On 11 November 1980 the Queen formally opened the Bankside Gallery. The opening exhibition was of Turner watercolours of the Thames, lent by the British Museum, together with works by Turner's contemporaries, from the RWS's permanent collection. This, though an artistic triumph, proved a very costly exercise, and coupled with the failure of the Bankside Gallery Trust to raise a great deal of money, resulted in financial difficulties for both Societies. It was thus never practicable to fulfil the hopes of holding further prestigious loan exhibitions and house international print and watercolour Biennales in alternate years. The print Biennale would have been in the year between the Bradford Print Biennales.

In April 1981 the Council agreed that there should be two RE shows a year, the spring exhibition was to be open, and the autumn to be of members only. The members' show was to include specially invited artists and each member was allowed to exhibit drawings alongside their prints. The two 1981 shows raised £15,128 in sales. Even allowing for inflation, this was encouraging. Nevertheless it was necessary to take the decision that year to increase the subscription once to £30. At the suggestion of Anthony Gross, a founder member of the highly regarded Parisian printmaking society, *La Jeune Gravure Contemporaine*, there was an exchange between the two societies. In June 1982 27 members of the RE held an exhibition at the Centre d'Art de la Rive Gauche and in the autumn of 1983 the French artists showed at the Bankside Gallery. The previous August there had been an exhibition of Hungarian painter–etchers. These were the first of a series of exhibitions of work by foreign printmakers.

Harry Eccleston was elected to a second term of office for another year in 1982, after a close contest with his Vice-President and friend, Charles Bartlett. His Presidentship continued to be extended on an annual basis until 1989. In 1983 Malcolm Fry retired as Director of the Bankside Gallery, after 20 years' association with the Society. He was replaced by his assistant, Michael Spender. The RE could no longer afford to pay half the expenses of the Gallery. It was decided that the Bankside Gallery Trust should be responsible for all gallery and society costs, and that the two Societies should pay agreed annual sums from their income for the services from the Gallery. In effect this step was to distance the two Societies from the running of the Gallery outside their own exhibitions, although each Society had a representative on the committee overseeing its running.

The RE started to participate at Art Fairs, taking a space at the Barbican International Contemporary Art Fair in January 1984 and at the Bath Contemporary Art Fair and the Olympia International Art Fair the following year. At the 1984 Annual General Meeting it was decided to raise the fee for existing members of the PCC to £15 from January 1985 and to £20 from January 1986, while new members would have to pay £20 at once. Members were to become automatically members of the newly formed Friends of Bankside Gallery. 1984 was one of the most successful years in terms of sales with £17,912.40 raised in the summer exhibition for which a special display of etchings and engravings over the past 25 years by Hayter was mounted, and a further £1,677.36 from the Christmas show. However sales fell back dramatically in 1985.

There was a growing realisation that the Society needed to have a wider appeal. David Fleming, an outsider from the field of public relations, in a letter of 13 August 1985 wrote, 'I think we are seriously handicapped by the name Royal Society of Painter-Etchers and Engravers. It gives the impression of ancients working on high stools under gas-lamp, breathing noxious fumes and producing stuff of no particular interest to the late Twentieth Century newly-wed. It is so long that there is a temptation to stop reading or listening before you get to the end of it. Hardly anyone knows what a painter-etcher or engraver is. In terms of useful information contained in the name, the Society might as well be called the Royal Society of Mumble–Jumble.

How about a change? Royal Society of Print-makers would do…'

In 1985 for the first time, thanks to the efforts of Lynne Williams, the annual show was sponsored, by Diners' International. Her hard work also brought about an impressive series of prizes largely from companies that supplied materials and tools to artists from Falkener Fine Paper, Barcham Green, Michael Putnam, Winsor and Newton, Clark Consultancy, Atlantis Paper, Whatman Paper, Clarendon Graphics, Galleries Magazine, Garton & Cooke, and C.C.A.Galleries.

The modernists finally won their battle, for the 1987 exhibition at long last admitted lithographs, but went far beyond this by accepting screenprints, computer prints and monoprints. This inclusivity meant that the Society needed to change its name. There was some debate over the most suitable title. Among the suggestions were Royal Artist Printmakers and Royal Engravers. The Society finally adopted the suggestion of Wilfred Fairclough. The long and costly process of changing the Royal Charter, the bye-laws and the name to the Royal Society of Painter–Printmakers, was not completed until 1991. The long accepted abbreviation for the title, RE, was retained. The decision to open the doors to all forms of printmaking led to an immediate response. No less than 1800 works were submitted for the 1987 show. At the following Annual General Meeting it was decided to double the number of Fellows from 50 to 100 in order to increase the opportunities for younger artists.

The Last Ten Years

On 23 October 1989 on the retirement of Harry Eccleston, the youngest ever and first non-British President, the American, Joseph Winkelman was elected, after a contest with the first woman candidate for the Presidency, Vikki Slowe. He was also the first President since the founder to have no association with a London College of Art, having trained at the Ruskin School of Art in Oxford. Fittingly, a century on from the inconclusive debates during Haden's time over original versus reproductive printmaking, Winkelman was much involved in discussions with Simon Brett of the Society of Wood Engravers on a committee set up to suggest a BSI *Standard for the Print*. The final draft, which was written by him with the assistance of Pat Gilmour, was published in *Print Quarterly,* XII, 1995, pp.179-181. The term 'original' was not included in the draft, largely because of its current use in some parts of the commercial world for prints that were totally reproductive.

Seven categories of prints were proposed to differentiate between the degree of the artist's actual involvement with the production of the print. The range extended from 'A', denoting that 'the artist alone created the matrix and made the impressions from it', to 'D', 'E','F' and 'G', where the artist had no direct involvement in the print's production neither preparing the matrix, nor pulling any impressions from it . 'B' covered cases where the artists made the matrix, but someone else printed the impressions taken from it. 'C' denoted the collaboration of another with the artist in the making of the matrix. 'D' denoted that both the preparation of the matrix and the printing was performed by others, but with the artist's supervision and approval. 'E' denoted that the print had been made from a pre-existing image with the artist or his agent's permission, but that was the limit of their participation in its production –and hence was a reproduction. 'F' covered prints made without the permission of either the artist or his agent. 'G' referred to re-strikes, that is re-printings from a matrix for a print in categories' 'A', 'B', or 'C', without the artist's knowledge or permission. It was intended that this standard should be freely available to all interested parties. Unfortunately the BSI was privatised during the discussions over the wording of the standard, and the newly constituted body imposed a charge for copies of the standard, even on those who had been involved with its formulation!

A special mezzotint exhibition was held in 1990 to coincide with the publication of Carol Wax's history, *The Art of Mezzotint*. This was organised by Peter Ford, a member, who also ran the innovative Off-Centre Gallery in Bristol. Ford, who had been involved with the organisation of a pioneering touring show of the works of members of the Society in Russia in 1989, was to introduce a series of fine Eastern European printmakers to British art lovers and to the Society, particularly through the 1992 exhibition of *A Time of Transition. Contemporary Printmaking from Russia and Ukraine,* shown at the Bankside Gallery in 1994. Among the artists to be elected to membership were the Pole, Piotr Ciesielski, the Ukrainian, Konstantin Kalinovich, and the Russians, Konstantin Chmutin and Pavel Makov. Ford was also one of the organisers of the *1st British International Miniature Print Exhibition* in 1994.

Winkelman made judicious use of Byelaw 30A to recruit new members and was very active in encouraging artists to submit for election. He also insisted that a third of the Council should retire each year, a rule that periodically had been neglected, in order to ensure a continual infusion of new blood into the running of the Society. This was a recurrent problem. Throughout the Society's history certain leading figures remained on the Council for extended periods. William Robins, for instance, was a member for 41 consecutive years, and in the postwar period, Stanley Dent and Joan Hassall for 26.

The membership of the Print Collectors' Club had been steadily declining in the 1980s. By 1989 it had fallen to 200. Jolyon Drury, son of Paul Drury, headed a working party to look into its future. The party recommended that, as the prints commissioned were not sufficiently prestigious to attract new members, in future a single print rather than three should be published each year and that the subscription fee should be raised once more to £75. This proposed trebling of the rate proved unacceptable at the Annual General Meeting. Instead a figure of £40 was decided on. However, the decision of the Inland Revenue that the Print Collectors' Club was not entitled to charitable status, and therefore liable for tax, including a large amount of back tax, forced the Society in 1992 to vote to amalgamate the financial, management and administrative affairs of the Club with those of the Friends of the Bankside Gallery. In place of the PCC, the Friends of the RE and RWS, from 1993, published a print annually.

The arrival in 1993 of Judy Dixey, as Director of the Bankside Gallery in place of Michael Spender, has proved significant in fostering improvements in the area of business management. Her background as an accountant with experience of regional arts organisations has proved invaluable to both societies in strengthening their financial management. With the help of a grant of £10,000 from the Wolfson Foundation, the Bankside Gallery was refurbished in 1990. However, the early 1990s proved tough years financially and the Bankside Trust failed to raise the sums of money that had been hoped for. One of the economies was the closing of the RE's journal that had been edited so well by Michael Blaker. Instead, the recently founded independent periodical, *Printmaking Today,* was adopted as the Society's official journal and Michael Blaker joined its board.

In 1995 David Carpanini, Professor of Art at the University of Wolverhampton, succeeded Winkelman, who had agreed in 1994 to serve as President for an extra year. He had trained at Gloucestershire College of Art and Design (formerly Cheltenham College of Art), where Stanley Dent RE had been Principal for nearly 25 years, the Royal

College of Art and the University of Reading. Gloucestershire College was one of group of art schools outside Central London that had become suppliers of printmakers to the Society. Fellows in these colleges, such as Wilfred Fairclough, Head of Department at Kingston since 1950, and his son, Michael Fairclough at the West Surrey College of Art at Farnham, were of vital importance in introducing artists to the Society. Recruits from London art schools were encouraged particularly by David Glück at the Central. Camberwell, the Slade, Wimbledon, Goldsmiths' and the Royal College all have provided new members. In recent years the Society has not stood still, initiating with the Federation of British Artists, the annual open Print Exhibition in the Mall Galleries and an exchange of exhibitions with the artists of the Manhattan Graphics Center in 1997 and 1998.

Recent developments in the vicinity of Bankside offer great opportunities to the Society at the start of the next century to expand knowledge and appreciation of the best in contemporary printmaking. The Tate Gallery is about to open its new establishment in the former Bankside Power Station which will be reached from the north bank of the Thames by the Millennium Bridge designed by Sir Norman Foster and Sir Antony Caro, thus facilitating access for art lovers working in the City. The Jubilee line extension too will ensure that there will be no difficulty in reaching a budding quarter for the visual arts. This part of Southwark already has attracted new galleries including Purdy Hicks and the Jerwood Space as well as the leading dealer in twentieth century artist's books and exhibition catalogues, Marcus Campbell. The former warehouse buildings in the area also offer studio space to an increasing community of practising artists. Plans are being drawn up for the improvement of the exhibition space in the Bankside Gallery, which should enable the Society and its sister, the Royal Society of Painters in Watercolours, to meet the exciting challenges that face them in the twenty-first century in good heart.

Bibliography

I am much indebted to Timothy Wilson and his colleagues in the Ashmolean Museum, Oxford, Judy Dixey and the staff of the Bankside Gallery, the staff of the Archives of American Art, the Centre for Whistler Studies, University of Glasgow, Glasgow University Library and New York Public Library for facilitating my research. I have benefited greatly from help from Charles Bartlett, Harry Eccleston, Simon Fenwick, Malcolm Fry, Ann Le Bas, Agatha Sorel, Clare Tilbury, Timothy Wilcox and Joseph Winkelman. Emma Chambers most generously allowed me to read the page proofs of the relevant chapters of her book on *The Etching Revival*. Ian Lowe read the manuscript and made some extremely helpful comments. Any errors, however, are entirely the responsibility of the author.

Principal Manuscript and Archival Sources

The Archives of the Royal Society of Painter-Printmakers, Bankside Gallery, London
Catalogue of the Diploma Etchings of the RSPE made by the Secretary in June 1894
Catalogues of exhibitions(some annotated) 1881-1998
Charter of incorporation and bye-laws 1911,1923,c.1927
Correspondence
Council Minute Books 1880-1986
Exhibitions Registers 1889-1933
Financial statements and reports 1881-1914
Memorandum of agreement of association 1881
Amended Memorandum of agreement of association 1887
Minute books of Print Collectors' Club 1921-1991

Newspaper cuttings book 1889-1910,1933-1969
Rules and regulations 1888,1891,1894,1899,1902
Sales books 1890-1942 , 1955-1968

Glasgow University Library, Special Collections–Sir
Francis Seymour Haden Papers

New York Public Library –Samuel Avery Papers

Ashmolean Museum, Oxford -Sir J.C.Robinson
Papers

Writings by Sir Francis Seymour Haden
About Etching, London, 1879
'The Art of the Painter-Etcher', *The Nineteenth Century,*
XXVII, 1890, pp.756-765; reprinted as *The Art of the
Painter-Etcher; What It Is- What It is Not, Being the First
Presidential Address of the Royal Society of Painter-Etchers
…,*London, 1890
'Mr Seymour Haden on Etching', *The Magazine of
Art,*II,1879,pp. 188-191,221-224,262-264(summary of
Haden's lectures delivered at the Royal Institution by
Marcus B.Huish)
'On the Revival of Mezzotint as a Painter's Art', *Harper's
Magazine,* LXX, 1885, pp.231-246.
*Presidential Address to the Royal Society of Painter-Etchers,
1901,* London, 1902
*The Relative Claims of Etching and Engraving to Rank as
Fine Arts, and to be represented as such in the Royal Academy
of Arts,* London, 1883
'The Royal Society of Painter-Etchers', *The Nineteenth
Century,* XXIX, 1891, pp.775-781, reprinted as *The Art
of the Painter-Etcher, Being the Second Presidential Address to
the Royal Society of Painter-Etchers,* London, 1891

Secondary sources
*The Journal of the Royal Society of Painter-Etchers and
Engravers,* later becoming *The Printmakers' Journal of the
Royal Society of Painter-Printmakers* and the publications of
The Print Collectors' Club
American Art Review, II, 1881
Dictionary of National Biography
Dudley Gallery, London, *Exhibition of Works in Black and
White,* 1876-1881
Fine Prints of the Year, I – XVI, 1923-1938
Sigrid Achenbach, *Berufskünstler und Amateure. Whistler-
Haden und die blüte der Graphik in England,* Staatliche
Museen Preussischer Kulturbesitz, Berlin, 1985
Anon, 'The Royal Painter-Etchers', *Art and Reason,* IV,
1938, pp.6-7.
Anon, 'The rules of The Print Sellers' Association', *The
Artists' Critical Record,* I, 1882, pp.172-173, 195-196 and
229-230.
Anon, 'The Society of Painter-Etchers', *The Artist's*

Record and Art Collectors' Guide, I, 15 November
1887, pp.105-106
Anon, 'The Royal Society of Painter-Etchers',
English Etchings, VIII, 1889-1891, pp.7-8, 36-38,67-
68
Hilary Beck, *Victorian Engravings,* Victoria and
Albert Museum, London, 1973
Emma Chambers, *'An Indolent and Blundering Art'?
The Etching Revival and the Redefinition of Etching
in England 1838-1892,* Scolar Press, Aldershot,
1999
Robin Garton, in Robin Garton ed., 'Etching 1855-
1935', *British Printmakers 1855-1955. A century of
printmaking from the Etching Revival to St Ives,* Garton
& Co in association with Scolar Press, Devizes,
1992, pp.125-127.
Robert H. Getscher, *The Stamp of Whistler,* Allen
Memorial Museum of Art, Oberlin, 1977
Richard T. Godfrey, *Printmaking in Britain. A
general history from its beginnings to the present day,*
Phaidon, Oxford, 1978
Basil Gray, *The English Print,* London, 1937
Kenneth M. Guichard, *British Etchers 1850-1940,*
London, 1981
Elton W. Hall, 'R.Swain Gifford and the New York
Etching Club', in David Tatham ed., *Prints and
Printmakers of New York State 1825-1940,* Syracuse
University Press, 1986
Martin Hardie, *Frederick Goulding, Master Printer of
copper plates,* Aeneas Mackay, Stirling, 1910
C.G. Holme, 'Youth and the Royal Society of
Painter-Etchers and Engravers', *The Studio,* CIII,
1932, pp.249-256
Simon Houfe, *Dictionary of British Book Illustrators
and Caricaturists 1800 -1914,* Antique Collectors'
Club, Woodbridge, 1978
Sidney C. Hutchison, *The History of the Royal
Academy 1768-1986,* Robert Royce Limited,
London, 1986
Pauline Winchester Inman, ' A History of the
Society of American Graphic Artists', *Artist's Proof,*
6, 3,1963-64 , pp.41-45
Anne Koval, *Responses to J.M.Whistler's Theory and
Practice: The Followers in Britain c.1880-1910,* Ph.D.,
Birkbeck College, London, 1997
Gladys Engel Lang and Kurt Lang, *Etched in Memory-
The Building and Survival of Artistic Reputation,* The
University of North Carolina Press, Chapel Hill and
London, 1990
James Laver*, A History of British and American
Etching,* London, 1929

James Laver, 'Seymour Haden & the Old Etching Club', *Bookman's Journal and Print Collector*, X, June 1924, pp.87-91

James Laver, ' Recent Engraving and Etching', *Artwork*, IV, 1928, pp.15-19

Katharine A. Lochnan, *The Etchings of James McNeill Whistler*, Yale University Press in association with The Art Gallery of Ontario, New Haven and London, 1984

Ian Lowe, 'The History of the RE', *Journal of the Royal Society of Painter- Etchers and Engravers*, XI, 1990, pp.49-53

Ian Lowe, 'The History of the RE Part 2', *Journal of the Royal Society of Painter-Etchers and Engravers*, XII, 1991, pp.47-48

E.S. Lumsden, *The Art of Etching*, London, 1924

Margaret F. MacDonald, 'Whistler and his followers', in Craig Hartley and Susan Ridyard ed., *The Print in Britain 1790-1930. A private collection*, Fitzwilliam Museum, Cambridge, 1985, pp.96 - 102

Ian Mackenzie, *British Prints*, Antique Collectors' Club, Woodbridge, 1987

Lauris Mason and Joan Ludman, *Print Reference Sources. A selected bibliography*, KTO Press, Milweed, 1981

Kemille Moore, *The Revival of Artistic Lithography in England 1890-1913*, Ph.D., University of Washington, University Microfilms, 1990

Sir Francis Newbolt, *The History of the Royal Society of Painter-Etchers & Engravers 1880-1930*, The Print Collectors' Club, London, 1930

F.M. Redgrave, *Richard Redgrave, C.B., R.A. A Memoir compiled from his diary*, Cassell & Company Limited, London, 1891

Timothy Riggs, *Index to Oeuvre-Catalogues of Prints by European and American Artists*, New York, 1983

H.R. Robertson, 'The Present Position of the Art of Etching', *English Etching*, VI, July 1886, pp.1-3

Malcolm C. Salaman, 'The Collector among the "Painter-Etchers", *The Bookman's Journal and Print Collector*, III, 1921, pp.270-273

Malcolm C. Salaman, 'Etchings and engravings from the 1935 Exhibition of the Royal Society of Painter – Etchers', *The Studio*, CIX, 1935, pp.194-199

Richard S. Schneiderman, *A Catalogue Raisonné of the prints of Sir Francis Seymour Haden*, Robin Garton Ltd, Devizes,1997(first edition 1983)

Frank Short, *Etchings and Engravings; What they are, and are not, with some notes on the care of prints*, Royal Society of Painters–Etchers and Engravers, London, 1912

Sir Frank Short, 'The R.E. Its aims and achievements', *The Studio*, CX, 1935, pp.129- 136

Walter Shaw Sparrow, *A Book of British Etching from Francis Barlow to Francis Seymour Haden*, John Lane, London, 1926

Michael Spender, 'Michael Spender, R.E. Secretary, analyses and compares Sales trends at the Exhibitions over the past decade', *The Journal of the Royal Society of Painter-Etchers and Engravers*, V, 1983, pp. 3-4

Allen Staley, 'Introduction', *The Stamp of Whistler*, Allen Memorial Museum of Art, Oberlin, 1977

Allen Staley ed., *The Post-Pre-Raphaelite Print*, Miriam & Ira D.Wallach Art Gallery, Columbia University, New York, 1995

E. Bonney Steyne, ' Painter-Etchers, Old and New', *Art Review*, I, 1890, pp.126-127

"Straight", ' 'The Painter-Etchers' Exhibition', *The Architectural Review*, XCV, June 1919

F.C.T., 'The Society of Painter-Etchers', *The Artist and Journal of Home Culture*, VI, 1 June 1885, pp.182-183

Martha P. Tedeschi, *How Prints Work*, Ph.D., Northwestern University, University Microfilms, 1994

Grant M.Waters, *Dictionary of British Artists Working 1900-1950*, Eastbourne Fine Arts, Eastbourne, 1975

Frederick Wedmore, *Etching in England*, George Bell, London, 1895

Frederick Wedmore, 'The Royal Society of Painter-Etchers', *The Studio*, V, 1895, pp.22-26

James McNeill Whistler, *The Gentle Art of Making Enemies*, William Heinemann, London, 1890

James McNeill Whistler, *The Piker Papers*, London, 1881

Joseph Winkelman, 'A British standard for prints', *Print Quarterly*, XII, 1995, pp. 179-181

For bibliographies on individual members of the Society see Garton, *British Printmakers 1855-1955*, Getscher, Houfe and Riggs(cited above), as well as the entries in standard dictionaries of artists such as Saur, Thieme-Becker, Vollmer and David Buckman, *The Dictionary of Artists in Britain since 1945*, Art Dictionaries Ltd., Bristol, 1998.

Fellows, Associates and Honorary Fellows of The Royal Society of Painter-Printmakers

PATRON (since 1985) His Royal Highness Prince Michael of Kent

Artists are listed in the chronological order in which they joined the Society. For alphabetical listing, see the Index of People on page 131. Originally all members were automatically Fellows, but from 1887 the status of Associate was introduced, and from that date those Associates who were subsequently elected to Fellowship are so listed in their entries. The RE number which precedes the title of a Diploma work is the catalogue number given by the Ashmolean Museum, where the prints are held in chronological sequence. Gaps in the numbers are due to a variety of reasons and do not indicate unlisted members or prints. Honorary members did not always give prints, and occasionally in the past the Society has failed to secure a Diploma print from a new member. Birth and death dates are given where known; otherwise information is given as available. Honours and titles listed are comprehensive rather than held at time of election.

ABBREVIATIONS

A.R.A.	Associate of the Royal Academy
A.R.E.	Associate of the Royal Society of Painter-Printmakers
A.R.S.A.	Associate of the Royal Scottish Academy
A.R.W.S.	Associate of the Royal Watercolour Society
F.S.A.	Fellow of the Society of Antiquaries
P.P.R.A.	Past President of the Royal Academy
P.P.R.E.	Past President of the Royal Society of Painter-Printmakers
P.P.R.W.S.	Past President of the Royal Watercolour Society
P.R.A.	President of the Royal Academy
P.R.E.	President of the Royal Society of Painter-Printmakers
P.R.I.	President of the Royal Institute of Painters in Watercolour
P.R.W.S.	President of the Royal Watercolour Society
R.A.	Royal Academician
R.B.S.	Fellow of the Royal Society of British Sculptors
R.E.	Fellow of the Royal Society of Painter-Printmakers
R.I.B.A.	Fellow of the Royal Institute of British Architects
R.S.A.	Royal Scottish Academician
R.S.W.	Member of the Royal Scottish Watercolour Society
R.W.S.	Member of the Royal Watercolour Society

Sir Francis Seymour Haden, M.D., P.R.E. (1818-1910)
Fellow and President 1880-1910
RE1 Wareham Bridge 1877
Drypoint 149 x 226 mm

Heywood Hardy, A.R.W.S. (1842-1933)
Fellow 1880
RE2 Horse's Head
Drypoint 323 x 241 mm

Prof. Sir Hubert von Herkomer, C.V.O., R.A., R.W.S. (1849-1914)
Fellow 1880-81, re-elected 1881 & 1892
RE3 Self Portrait with his Children 1879
Etching 352 x 204 mm
RE39 Portrait of W.H.Thompson D.D. 1881
Mezzotint 496 x 380 mm
RE155 The Witches' Pet 1891
Drypoint 157 x 150 mm
RE156 Gwenddydd 1891
Etching 142 x 95 mm

Prof. Alphonse Legros (1837-1911)
Fellow 1880-85, re-elected 1895
RE4 Death and the Woodman c.1875
Etching 373 x 270 mm

RE5 Portrait of Francis Seymour Haden 1881
Mezzotint 247 x 172 mm
RE170-2 Triomphe de la Mort
Etchings & drypoint 190 x 138; 214 x 415;
320 x 486 mm

Robert Walter Macbeth, R.A., R.W.S. (1848-1910)
Fellow 1880-87, re-elected 1892
RE6 Coming from St Ives 1878
RE158 His Last Copper
Etchings 176 x 312; 147 x 123 mm

James Joseph Jacques Tissot (1836-1902)
Fellow 1880
RE7 An Uninteresting Story 1878
Etching & drypoint 314 x 203 mm

Philip Gilbert Hamerton (1834-1894)
Hon. R.E. 1880
RE8 Hôtel de Beauchamp
Etching 266 x 187 mm

Edward Hamilton, M.D. (1834-1893)
Hon. R.E. 1881, Hon. Treasurer 1880-90

Sir William Richard Drake, F.S.A. (1817-1890)
Hon. R.E., Hon. Secretary 1881-90

Richard Fisher, F.S.A. (d.1890)
Fellow 1881, Hon. Curator 1881-90

Frank Holl, R.A., A.R.W.S. (1845-1888)
Fellow 1881, Hon. Fellow 1886
RE9 Sketch of Heywood Hardy
Drypoint 378 x 296 mm

James Clarke Hook, R.A. (1819-1907)
Fellow 1881, Hon. Fellow 1886
RE10 Mussel Gatherers 1879
Etching 173 x 237 mm

Colin Hunter, R.S.W. (1841-1904)
Fellow 1881
RE11 The Gareloch
Etching 302 x 576 mm

Charles Samuel Keene (1823-1891)
Fellow 1881
RE12 Walberswick Pier 1867
Etching & drypoint 103 x 161 mm

William Ewart Lockhart, A.R.W.S., R.S.A.
(1846-1900)
Fellow 1881

Briton Riviere, R.A. (1840-1920)
Fellow 1881
RE13 The King Drinks
Etching 271 x 416 mm

Charles West Cope, R.A. (1811-1890)
Fellow 1881, Hon. Fellow 1886
RE14 Life School, Royal Academy 1865
Etching 154 x 248 mm

Sir Ernest George, R.A. (1839-1922)
Fellow 1881
RE15 Houses on the Ill, Strasbourg
Etching 192 x 272 mm

John Evan Hodgson, R.A. (1831-1895)
Fellow 1881, Hon. Fellow 1886
RE16 Artists and Amateurs 1878
Etching 185 x 244 mm

Henry Stacey Marks, R.A., R.W.S. (1829-1898)
Fellow 1881, Hon. Fellow 1886
RE17 Rustic Wonder 1859
Etching 138 x 93 mm

**Sir Edward John Poynter, Bt., G.C.V.O.,
P.R.A., R.W.S.** (1836-1919)
Fellow 1881, Hon. Fellow 1886
RE18 Study 1881
Etching 277 x 205 mm

**Sir Lawrence Alma-Tadema, O.M., R.A.,
R.W.S.** (1836-1912)
Fellow 1881, Hon. Fellow 1886
RE19 Tesselschade at Alkmaar 1879
Etching 161 x 123 mm

Otto Henry Bacher (1856-1909)
Fellow 1881
RE20 Market, Florence 1880
Etching & drypoint 181 x 257 mm

Oliver Baker (1856-1939)
Fellow 1881
RE21 In Chancery
Etching 155 x 231 mm

Auguste Ballin (b.1842, still active 1899)
Fellow 1881
RE22 Putting in Commission - Marine 1650 1879
Etching 202 x 359 mm

Henry William Batley (fl.1873-1893)
Fellow 1881
RE23 Book Frontispiece 1880
Engraving 283 x 158 mm

Albert Fitch Bellows (1830-1883)
Fellow 1881
RE24 Stratford Street, Connecticut
Etching 120 x 180 mm

Félix Buhot (1847-1898)
Fellow 1881
RE25 La Place Pigalle Matin d'Été en 1878
Etching & aquatint 260 x 348 mm

Richard Samuel Chattock (1825-1906)
Fellow 1881-87, re-elected 1891
RE26 The Moorhen
RE144 Fittleworth Mill
Etchings 175 x 353; 182 x 274 mm

Frederick Stuart Church (1842-1924)
Fellow 1881
RE27 The Mermaid 1880
Drypoint 222 x 147 mm

Theodore Irving Dalgliesh (1855-1941)
Fellow 1881
RE28 In the Gloaming
Etching 227 x 139 mm

Frank Duveneck (1848-1919)
Fellow 1881
RE29 The Riva, looking towards the Caserna 1880
Etching 249 x 374 mm

John M. Falconer (1820-1903)
Fellow 1881
RE30 Home no Longer 1880
Etching 119 x 182 mm

Henry Farrer (1843-1903)
Fellow 1881
RE31 In New York Harbour 1879
Etching 204 x 304 mm

George Straton Ferrier (d.1912)
Fellow 1881
RE32 Herring Fleet leaving Wick Harbour 1879
Etching 198 x 301 mm

Robert Swain Gifford (1840-1905)
Fellow 1881
RE33 Padanaram Salt Works 1878
Etching 168 x 340 mm

Charles Storm van 's Gravesande (1841-1924)
Fellow 1881
RE34 The Old Pier at Flushing
Drypoint 229 x 305 mm

Axel Herman Haig (1835-1921)
Fellow 1881
RE35 In a Swedish Country Church 1881
RE36 Halle der Schiffergesellschaft, Lübeck 1880
Etchings 253 x 176; 446 x 302 mm;

A. Brames Hall (fl.1881-1883)
Fellow 1881
RE37 Japanese Bronze Incense-burner 1881
Etching 317 x 218 mm

Howard Helmick (1845-1907)
Fellow 1881
RE38 Thomas Carlyle
Etching 249 x 201 mm

John Postle Heseltine (1843-1929)
Fellow 1881
RE40 Great Yarmouth 1876
Etching 186 x 298 mm

Thomas Huson (1844-1920)
Fellow 1881
RE41 Drizzle 1880
Mezzotint & etching 176 x 266 mm

George Percy Jacomb-Hood (1857-1929)
Fellow 1881
RE42 "A gentle knight was pricking on the Plaine"
RE43 Italian Peasant
Etchings 200 x 155; 278 x 209 mm

John W. Buxton Knight (1843-1908)
Fellow 1881
RE44 Plighting Troth

RE45 Peasants in a Landscape
RE46 Landscape with Sheep
RE47 Landscape with River and Bridge
Etchings 316 x 220; 202 x 302; 174 x 253;
174 x 253 mm

David Law (1831-1901)
Fellow 1881
RE48 On the Orchy
Etching 274 x 399 mm

C.W. Mansel Lewis (1846-1931)
Fellow 1881
RE49 A Welsh Hat
Drypoint 251 x 201 mm

Otto Theodore Leyde, R.S.W. (1835-1897)
Fellow 1881
RE50 Salmon Fishing in the Firth of Forth
Drypoint 171 x 280 mm

Léon Lhermitte (1844-1925)
Fellow 1881
RE51 La Boucherie
Etching 175 x 165 mm

John MacWhirter, R.A. (1839-1911)
Fellow 1881
RE52 A Fisherman's Haven
Etching 184 x 263 mm

Herbert Menzies Marshall, R.W.S. (1841-1913)
Fellow 1881
RE53 At Rotherhithe 1881
Etching 226 x 148 mm

Mortimer Menpes (1860-1938)
Fellow 1881
RE54 Old Corner, Vitré 1881
Etching 250 x 164 mm

Hans Meyer (1846-1919)
Fellow 1881
RE55 Head of an old Man 1881
Etching 269 x 212 mm

Mary Nimmo Moran (1842-1899)
Fellow 1881
RE56 Goose Pond
Etching 177 x 229 mm

Thomas Moran (1837-1926)
Fellow 1881
RE57 The Rainbow
Etching 95 x 196 mm

Charles Oliver Murray (1842-1924)
Fellow 1881
RE58 Left Lonely
Etching 312 x 267 mm

Stephen Parrish (1846-1938)
Fellow 1881
RE59 Fishermen's Houses, Cape Ann 1849
Etching 302 x 480 mm

Léon Richeton (1854-1934)
Fellow 1881
RE60 Portrait of Dean Stanley
Etching 329 x 248 mm

Edouard Rischgitz (1828-1909)
Fellow 1881
RE61 Le Grénouille et le Boeuf
Etching 239 x 178 mm

Charles Paul Renouard (1845-1924)
Fellow 1881
RE62 Une Visite sur les Toits de l'Opéra à Paris
c.1880-81
Etching & aquatint 222 x 193 mm

Henry Robert Robertson (1839-1921)
Fellow 1881
RE63 Rush Harvest
Etching 174 x 301 mm

William Scott (fl. 1880-1897)
Fellow 1881
RE64 Courtyard, S. Cosimato, Rome
Etching 214 x 146 mm

Charles William Sherborn (1831-1912)
Fellow 1881
RE65 Portrait of Francis Seymour Haden 1880
Engraving 149 x 86 mm

Charles Philip Slocombe (1832-1895)
Fellow 1881
RE66 In the Woods, Heidelberg 1879
Etching 378 x 224 mm

Frederick Albert Slocombe (1847-1920)
Fellow 1881
RE67 Pinner Hill 1880
Etching 374 x 234 mm

James David Smillie (1833-1909)
Fellow 1881
RE68 Up the Hill 1879
Etching 202 x 126 mm

William Spread (d.1909)
Fellow 1881
RE69 A Shop in Dinan, Brittany 1879
Etching 80 x 137 mm

George Stevenson (fl.1881-1888)
Fellow 1881
RE70 Derelict - Dawn
Etching 175 x 199 mm

William Strang, R.A. (1859-1921)
Fellow 1881
RE71 Poverty
RE72 Basket Carriers
Etchings 150 x 197; 230 x 159 mm

Dorothy Tennant (1855-1926)
Fellow 1881
RE73 Young Bacchanals at Play
Etching 82 x 122 mm

Percy Thomas (1845-1922)
Fellow 1881
RE74 Using the old Pump
Etching 118 x 82 mm

Robert Kent Thomas (1816-1884)
Fellow 1881
RE75 St. Cuthbert's Screen, St. Alban's Cathedral
Etching 207 x 151 mm

William Henry Urwick (1864-1943)
Fellow 1881
RE76 An old Thames Fisherman
RE77 Mountain Landscape
Etchings 136 x 236; 146 x 259 mm

John Watkins (d.1908)
Fellow 1881
RE78 Bronze Bust of Pope Alexander VII, by Bernini
Etching 245 x 169 mm

Charles John Watson (1846-1927)
Fellow 1881
RE79 London Bridge 1878
Etching 283 x 190 mm

Otto Weber, A.R.W.S. (1832-1888)
Fellow 1881
RE80 Mid-day Meal
Etching 232 x 452 mm

Samuel Henry Baker (1824-1909)
Fellow 1882
RE81 Mill on the Trystion, North Wales
Etching 149 x 219 mm

Wilfrid Williams Ball (1853-1917)
Fellow 1882
RE82 Stranded - Rye, Sussex
Etching 100 x 238 mm

Louis Alfred Brunet-Desbaines (1845-1939)
Fellow 1882
RE83 Sainte Chapelle, Paris
Etching 194 x 142 mm

Charles Edward Holloway (1838-1897)
Fellow 1882

RE84 Old Chelsea 1882
Etching 259 x 400 mm

Catherine Maude Nichols (1847-1923)
Fellow 1882
RE85 Fir Trees 1881
Drypoint 251 x 176 mm

Joseph Pennell (1857-1926)
Fellow 1882
RE86 Mommi Sauerkraut's Row, Philadelphia 1881
Etching 214 x 301 mm

Charles Adams Platt (1861-1933)
Fellow 1882
RE86A Portland, on the St. John 1882
Etching 280 x 523 mm

François Auguste René Rodin (1840-1917)
Fellow 1882
RE87 A Sketch 1881
Etching 200 x 248 mm

Ned (Edward) Swain (1847-1909)
Fellow 1882
RE88 Spur's Holt Orchard
Etching 150 x 199 mm

Hendrik Dirk Kruseman Van Elten (1829-1904)
Fellow 1882
RE89 On the Housatonic River
Etching 124 x 211 mm

Theodore M. Wendel (1857- after 1933)
Fellow 1882
RE90 Venice
Drypoint 242 x 192 mm

Walter William Burgess (1856-1908)
Fellow 1883
RE91 Cathedral of Limburg-on-the-Lahn 1883
Etching 424 x 310 mm

Joseph Knight (1837-1909)
Fellow 1883-88, re-elected 1892
RE92 Near Barmouth 1882
RE151 Trees, Evening
Mezzotints 338 x 296; 288 x 229 mm

William Holmes May (1839-c.1920)
Fellow 1883
RE93 The Avenue at Haddon
Etching 175 x 238 mm

George Woolliscroft Rhead (1854-1920)
Fellow 1883
RE94 Head of an old Man
Mezzotint & etching 379 x 294 mm

Edward Slocombe (1850-1915)
Fellow 1883

RE95 A dead Oak – New Forest, Hants 1885
Etching 353 x 223 mm

Stanley Berkeley (1855-1909)
Fellow 1884
RE96 Homeward wending in Silver Sheen 1883
Etching 147 x 314 mm

Edith Berkeley (fl.1880-1897)
Fellow 1884
RE97 Playmates
Etching 199 x 148 mm

Countess Feodora Gleichen (1861-1922)
Fellow 1884, Hon. Fellow 1891
RE98 Study of a Head
Etching 199 x 148 mm

Edgar Barclay (1842-1913)
Fellow 1885
RE99 Little Ones First
Etching 335 x 176 mm

H.J. Angley (fl.1883-1887)
Fellow 1885
RE100 Eynesford 1887
Etching 315 x 610 mm

Elizabeth Adela Armstrong (1859-1912)
Fellow 1885
RE101 Saying Grace
Etching 121 x 202 mm

Herbert Thomas Dicksee (1862-1942)
Fellow 1885
RE102 Hercules wrestling with Death for
the Body of Alcestes
Etching & aquatint 414 x 308 mm

Sir Alfred East, R.A. (1849-1913)
Fellow 1885
RE103 Moonlight
Mezzotint & etching 170 x 267 mm

Thomas Charles Farrer (1839-1891)
Fellow 1885
RE104 Going to the Lido 1885
Etching 193 x 295 mm

Alexis Forel (b.1852, still active 1889)
Fellow 1885
RE105 Old House and Church of St. Julien
Etching & aquatint 238 x 190 mm

Elinor Hallé (fl.1882-1914)
Fellow 1885
RE106 Miserere Nobis
Etching 290 x 202 mm

William Brassey Hole, R.S.A., R.S.W. (1846-1917)
Fellow 1885
RE107 Sir Alexander Grant 1893
Etching 160 x 110 mm

Sir Charles Holroyd (1861-1917)
Fellow 1885
RE108 Death of Torrigiano
RE109 Portrait of Alphonse Legros
Etchings 175 x 231; 251 x 175 mm

Charles Robertson, R.W.S. (1844-1891)
Fellow 1885
RE110 Yarmouth Fishing Boats 1885
Etching 470 x 665 mm

George Roller (b.1858)
Fellow 1885
RE111 The Last Straw 1883
Etching 245 x 162 mm

Sir Francis Job Short, P.R.E., R.A. (1857-1945)
Fellow 1885, President 1910-38
RE112 The Tide ebbs, Putney Bridge 1885
Mezzotint 125 x 181 mm

Richard Toovey (1861-1927)
Fellow 1885
RE113 A Dutch Harbour 1885
Etching 200 x 298 mm

George Gascoyne (1862-1933)
Fellow 1885, Hon. Fellow 1922
RE114 The Wanderer 1885
Etching 76 x 76 mm

Harrington Mann (1864-1937)
Fellow 1885
RE115 A Portrait
Etching 198 x 149 mm

Colonel Robert Charles Goff (1837-1922)
Fellow 1887
RE116 London Bridge
Etching 118 x 163 mm

Vereker Monteith Hamilton (1856-1931)
Fellow 1887
RE117 Head of a Girl
Etching 202 x 157 mm

Walter Richard Sickert, R.A. (1860-1942)
Fellow 1887-92, re-elected Associate 1925
RE118 Louie 1884
Etching 228 x 225 mm

Charles Frederick Allbon (1856-1926)
Associate 1887
RE119 Homeward 1888
Etching & drypoint 240 x 670 mm

Thomas Barrett (1845?-1924)
Associate 1887
RE120 Fishing Boats set Sail 1886
Etching 167 x 256 mm

Lillian Hamilton (b.1865, still active 1932)
Associate 1887
RE121 The Forge
Etching 226 x 254 mm

Alberto Maso Gilli (1840-1894)
Associate 1887

Victoria S. Hine (1840-1926)
Associate 1887
RE122 Durham Cathedral 1887
Etching 371 x 475 mm

Anna Lea Merritt (1844-1930)
Associate 1887-92, re-elected 1905
RE123 Eve 1887
RE226 Grandmother's Boa
Etchings & drypoint 332 x 480; 227 x 175 mm

Lawrence Barnett Phillips (1842-1922)
Associate 1887
RE124 Shipping
Etching 265 x 375 mm

Alexander Wallace Rimington (1854-1918)
Associate 1887
RE125 Old Zurich
Etching 196 x 323 mm

Gerald Robinson (b.1858, still active 1901)
Associate 1887
RE126 Portrait of Sir Francis Seymour Haden
Mezzotint 151 x 121 mm

Percy Robertson (1868-1934)
Associate 1887, Fellow 1908
RE127 Whitby
Etching 326 x 216 mm

Tristram Ellis (1844-1922)
Associate 1887
RE128 The "Victory" at Portsmouth 1888
Etching 316 x 456 mm

John Finnie (1829-1907)
Associate 1887, Fellow 1895
RE129 "Touch the hills and they shall smoke"
Mezzotint 246 x 399 mm

William Niven (active 1853, d.1921)
Associate 1887
RE130 Interior of St. Mildred, Bread Street
Engraving 266 x 204 mm

Hugh Paton (1853-1927)
Associate 1887
RE131 The Haven under the Hill
Etching 214 x 302 mm

Arthur C.H. Luxmoore (active 1880, d.1891)
Associate 1888
RE132 Doorway of Rochester Cathedral 1888
Etching & aquatint 175 x 126 mm

Arthur Robertson (1850-1911)
Associate 1888
RE133 The Last Romance
Etching 343 x 270 mm

**Sir David Young Cameron, R.A., R.W.S.,
R.S.A., R.S.W.** (1865-1945)
Associate 1889, Fellow 1895
RE134 Old Age 1891
RE135 A Perthshire Village
Etchings 265 x 158; 127 x 198 mm

Alfred William Strutt (1856-1924)
Associate 1889
RE136 A Return Visit 1889
Etching 230 x 413 mm

Francis S. Walker (1848-1916)
Associate 1890, Fellow 1897
RE137 Windsor Castle
Etching 120 x 189 mm

Alfred Walter Bayes (1831-1909)
Associate 1890, Fellow 1900
RE138 Old Houses in Camden Town
Etching 295 x 189 mm

William Henry Boucher (active 1890, d.1906)
Associate 1890
RE139 Full Length Portrait of a Lady
Etching 162 x 118 mm

Charles F. Robinson (fl.1874-1915)
Associate 1890
RE140 Rainham
Etching 171 x 251 mm

Félix Henri Bracquemond (1833-1914)
Hon. R.E. 1891
RE141 The Mole Catcher 1854
Etching 250 x 190 mm

Robert Bryden (1865-1939)
Associate 1891, Fellow 1899
RE142 The Corridor of The Art Schools, South
Kensington 1889
Etching 259 x 163 mm
RE143 The Weaver 1891
Drypoint 149 x 225 mm

Arthur Evershed, M.D. (1836-1919)
Associate 1891, Fellow 1898
RE145 At Sandwich 1882
Etching 102 x 224 mm

Oliver Hall, R.A., R.W.S., R.S.A. (1869-1957)
Associate 1891, Fellow 1895
RE146 Study of Trees 1889
Etching 119 x 162 mm

Charles Haslewood Shannon, R.A. (1863-1937)
Associate 1891

Axel Tallberg (1860-1928)
Associate 1891
RE147 A Farm-hand 1893
Etching 139 x 99 mm

F. Inigo Thomas (fl.1891-1929)
Associate 1891
RE148 Kaiserberg
Etching 246 x 165 mm

Frank Sternberg (b.1858, still active 1904)
Associate 1892

Charles E. Baskett (active 1883, d.c1929)
Associate 1892
RE149 Donyland Wood, Winter
Etching 114 x 160 mm

Edward William Charlton (1859-1935)
Associate 1892, Fellow 1907
RE150 Dinghy a'hoy!
Etching 144 x 217 mm

Ethel King Martyn (1863-1946)
Associate 1892, Fellow 1902
RE152 An Idyll
Etching 75 x 152 mm

Elizabeth Piper (fl.1892, d.1956)
Associate 1892
RE153 Musée de Cluny
Etching 269 x 160 mm

Charles Bird (fl.1892-1907)
Associate 1892
RE154 Doorway, St. Austin's, Bristol 1885
Etching 454 x 318 mm

Paul César Helleu (1859-1927)
Associate 1892, Fellow 1897
RE157 Ellen Helleu
Drypoint 323 x 299 mm

W.T.B. Roberts (fl.1880, d.1896)
Associate 1892
RE159 Coffee Tavern 1891
Etching 212 x 268 mm

Henri Béraldi (1848/9-1931)
Hon. R.E. 1892

Frank Laing (1862-1908)
Associate 1892
RE160 Pont de Stocades, Paris
Etching 141 x 225 mm

Susan Fletcher Crawford (1864-1918)
Associate 1893
RE161 Potato Diggers
Etching & drypoint 258 x 384 mm

Jessie Harrison (fl.1893-1936)
Associate 1893
RE162 The Mumbles, Swansea
Etching 124 x 173 mm

Hubert Schröder (1867-1940)
Associate 1893
RE163 Old Shoreham Bridge
Etching 105 x 170 mm

Annie Williams (fl.1887-1916)
Associate 1893
RE164 Charterhouse
Etching 158 x 217 mm

Minna Bolingbroke (fl.1888, d.1939)
Associate 1893, Fellow 1899
RE165 The Loom 1889
Drypoint 239 x 150 mm

Sir John Charles Robinson, K.C.V.O.
(1824-1913)
Fellow 1893
RE166 A Swollen Burn at Shandon
Etching & drypoint 169 x 245 mm

Ferdinand Boberg (1860-1946)
Associate 1894
RE167 Snowstorm 1899
Etching 381 x 232 mm

George W. Eve (1855-1914)
Associate 1894, Fellow 1903
RE168 Bookplate
Engraving 69 x 48 mm

Alfred Hartley (1855-1933)
Associate 1894, Fellow 1897
RE169 Hen and Chicks
Etching 201 x 150 mm

Henry Gibbs Massey (1860-1934)
Associate 1894
RE173 Runswick, Yorks
Etching 151 x 202 mm

Henry Macbeth-Raeburn, R.A. (1860-1947)
Associate 1894-99, re-elected Fellow 1920
RE174 The Alhambra 1893
Etching 165 x 247 mm

William Monk (1863-1937)
Associate 1894, Fellow 1899
RE175 The Life School 1894
Etching 349 x 438 mm

Constance Mary Pott (1862-1957)
Associate 1894, Fellow 1898
RE176 Southampton Water
Etching 125 x 175 mm

Ernest Stamp (1869-1942)
Associate 1894
RE177 Blind Beggar 1894
Drypoint 151 x 100 mm

Louis Monziès (b.1849, still active 1896)
Associate 1894

Frederick Vango Burridge (1869-1945)
Associate 1895, Fellow 1898
RE178 In the Orchard 1894
Etching 126 x 175 mm

Woodbine K. Hinchcliff (fl.1895-1913)
Associate 1895
RE179 Mrs Webber 1894
Etching 175 x 127 mm

Wilfred Thompson (fl.1884-1901)
Associate 1895
RE180 Mors Consolatrix
Etching 300 x 250 mm

Walter Stearne Hale (1869-1917)
Associate 1895
RE181 Corner of Old Philadelphia 1891
Etching 234 x 188 mm

James G. Murray (1861-1906)
Associate 1895
RE182 The Ingle Neuk
Etching 86 x 184 mm

Colin R. Carroll (active 1895, d.1900)
Associate 1895
RE183 Waterfall
Mezzotint 292 x 196 mm

Henry Robertson (fl.1885-1898)
Associate 1896
RE184 In the Springtime 1894
Etching 115 x 159 mm

Constance Gertrude Copeman (1864-1953)
Associate 1897

RE185 A Young Artist
Etching 163 x 118 mm

Gertrude Ellen Hayes (1872-1956)
Associate 1897
RE186 Street at Rouen
Etching 175 x 125 mm

Bernard Schumacher (b.1872, stilll active 1903)
Associate 1897
RE187 Old Hildesheim, Hanover
Etching 178 x 126 mm

Robert Spence (1871-1964)
Associate 1897
RE188 George Fox and the Scholars
Etching 305 x 229 mm

Sir Frederick Wedmore (1844-1921)
Hon. R.E. 1897

Louise, H.R.H. Princess, Duchess of Argyll, Hon. R.W.S. (1848-1939)
Hon. R.E. 1897

Rogelio de Egusquiza (1845-1915)
Associate 1898
RE189 From Wagner's "Parsifal" 1894
Etching & drypoint 473 x 350 mm

Alfred Hugh Fisher (1867-1945)
Associate 1898
RE190 Gray's Inn
Etching 214 x 152 mm

John Park (active 1885, d.1919)
Associate 1898
RE191 North Shields Harbour 1897
Etching & drypoint 224 x 369 mm

Edward Millington Synge (1860-1913)
Associate 1898
RE192 Zaandam
Etching 200 x 149 mm

Reginald Edgar James Bush (1869-1956)
Associate 1898, Fellow 1917
RE193 On the Arno, Florence 1894
Etching 175 x 125 mm

William Henry Milnes (1865-1957)
Associate 1899
RE194 An Old Barn
Etching 103 x 149 mm

Adolph Campbell Meyer (1866-1919)
Associate 1899
RE195 Evening Calm
Mezzotint 199 x 153 mm

Sir Francis George Newbolt, K.C. (1863-1940)
Associate 1899
RE196 By Overstrand Hall 1899
Etching 166 x 116 mm

Charles Algernon Tomkins (fl.1881-1899)
Associate 1899
RE197 Teesmouth
Mezzotint 222 x 308 mm

John Wright (1857-1933)
Associate 1899, Fellow 1917
RE198 The Pond
Etching 152 x 202 mm

Millicent Bramley-Moore (fl.1899-1914)
Associate 1899
RE199 An Interior 1898
Etching 177 x 167 mm

Eugène Béjot (1865-1931)
Associate 1899, Fellow 1908
RE200 Le Pont Sully, Paris 1919
Etching 246 x 215 mm

Edgar Chahine (1874-1947)
Associate 1900
RE201 Au Château Rouge 1899
Drypoint 280 x 334 mm

Elisabeth Decker (fl.1889-1904)
Associate 1900
RE202 Harvest 1899
Etching 124 x 175 mm

Amelia Bauerlë (active 1897, d.1916)
Associate 1900
RE203 A Song of the Sea
Engraving 123 x 175 mm

Mary Annie Sloane (1868-1961)
Associate 1900
RE204 A Weaver 1892
Etching 253 x 309 mm

Henri B. van Raalte (1881-1928)
Associate 1900
RE205 The Boat Builder's Shop at Rye 1900
Etching 258 x 354 mm

Margaret Kemp-Welch (active 1898, d.1968)
Associate 1901
RE206 The Cross Roads
Etching 160 x 174 mm

E. Philip Pimlott (fl. 1893-1940)
Associate 1901
RE207 Building of Boats, St. Ives 1901
Etching 235 x 162 mm

David Waterson (1870-1954)
Associate 1901, Fellow 1910
RE208 The Mill
Etching & aquatint 177 x 239 mm

Alfred Kedington Morgan (1868-1928)
Associate 1901
RE209 A Corner in Bury, Sussex
Etching 175 x 249 mm

Bertha Gorst (b.1873, still active 1929)
Associate 1902
RE210 Evening
Etching 152 x 212 mm

Mary E. Kershaw (1884-1968)
Associate 1902
RE211 Girl Feeding Fowls
Etching 189 x 138 mm

Hermann Struck (1876-1944)
Associate 1902

Luke Taylor (1876-1916)
Associate 1902, Fellow 1910
RE213 Old Chelsea
Etching 145 x 195 mm

Percy Wadham (fl.1893-1905)
Associate 1902
RE214 Courtyard of the Hôtel de Gondi, Villeneuve-les-Avignon 1901
Etching 182 x 261 mm

Frank Willis (1865-1932)
Associate 1902
RE215 Grandfather's Tale 1902
Etching 91 x 135 mm

John Alexander Ness (d.1931)
Associate 1903
RE216 Muto's Lane, St. Andrews 1902
Etching 269 x 110 mm

William Lionel Wyllie, R.A. (1851-1931)
Associate 1903, Fellow 1903
RE217 Unloading a Leaky Timber Ship
Etching 231 x 354 mm

Hedley Fitton (1859-1929)
Associate 1903, Fellow 1908
RE218 St. Martin's Church, London 1901
Etching 308 x 216 mm

Johann Nordhagen (1856-1956)
Associate 1903
RE219 A Viking 1903
Etching 254 x 195 mm

Gustave Bourcard (d.1925)
Hon. R.E. 1903

Sir Frank Brangwyn, R.A., A.R.W.S., R.S.W. (1867-1956)
Associate 1903 Fellow 1903

Adeline S. Illingworth (1868-1942)
Associate 1904
RE220 Courtyard of the Architect's House at Rothenburg 1902
Etching 203 x 254 mm

Hjalmar Molin (b.1868, still active 1928)
Associate 1904
RE221 Cathedral in Barcelona 1903
Etching & aquatint 555 x 398 mm

Georges Jeanniot (1848-1934)
Associate 1904
RE222 Fantassin Allemand
Drypoint 214 x 131 mm

Edward Julius Detmold (1883-1957)
Associate 1905
RE223 Study of a Tree
Etching 281 x 194 mm

Charles Maurice Detmold (1883-1908)
Associate 1905
RE224 The Dragon
Etching 166 x 274 mm

Sydney Lee, R.A., R.W.S. (1866-1949)
Associate 1905, Fellow 1915
RE225 On the Medway 1902
Etching 156 x 236 mm

Harold Percival (1868-1914)
Associate 1905
RE227 Marsh Hay 1904
Drypoint & aquatint 176 x 253 mm

Nathaniel Sparks (1880-1956)
Associate 1905, Fellow 1910
RE228 A Native of Sierra Leone
Etching 160 x 127 mm

James Robert Granville Exley (1878-1967)
Associate 1905, Fellow 1923
RE229 Houdan Fowls 1905
Etching 115 x 136 mm

Malcolm Osborne, C.B.E., P.R.E., R.A. (1880-1963)
Associate 1905, Fellow 1909, President 1938-62
RE230 Maggie, Study of a Girl's Head 1905
Etching 127 x 65 mm

Douglas Ian Smart (1876-1970)
Associate 1905, Fellow 1922
RE231 Old Houses at Walham Green
Etching 200 x 273 mm

Ethel Stewart (active 1904, d.1924)
Associate 1906
RE232 Evening Tide at Bridgwater
Etching 163 x 208 mm

Winifred Marie Louise Austen (1876-1964)
Associate 1907, Fellow 1922
RE233 The White Heron
Etching 214 x 105 mm

Martin Hardie, C.B.E., Hon. R.S.W. (1875-1952)
Associate 1907, Fellow 1920
RE234 Dulieu's Pig Farm 1907
Etching 160 x 200 mm

Herman Armour Webster (1878-1970)
Associate 1907, Fellow 1914
RE235 La Rue de la Parchemanerie, Paris 1907
Etching 274 x 174 mm

Eli Marsden Wilson (1877-1965)
Associate 1907
RE236 The Market, Ossett
Etching 101 x 175 mm

Anna Airy (1882-1964)
Associate 1908, Fellow 1914
RE237 September, 1907 1908
Etching 200 x 164 mm

Gustave Leheutre (1861-1932)
Associate 1908
RE238 Le Canal d'Eu
Etching 165 x 244 mm

Alfred Bentley, M.C. (1879-1923)
Associate 1908, Fellow 1913
RE239 Shipbuilding 1908
Etching 228 x 263 mm

George Percival Gaskell (1868-1934)
Associate 1908, Fellow 1911
RE240 The Castle of Neuschwanstein, Bavaria
Etching 223 x 250 mm

Mabel Catherine Robinson (1875-1953)
Associate 1908
RE241 The Old Smithy 1908
Etching 261 x 355 mm

Eleanor Fell (active 1899, d.1946)
Associate 1908
RE242 Lewes Castle
Aquatint 196 x 245 mm

William Henry Ansell (1872-1959)
Associate 1909
RE243 The Court of L'Ange Gabriel, St. Riquier 1908
Etching 227 x 175 mm

Ernest Stephen Lumsden, R.S.A. (1883-1948)
Associate 1909, Fellow 1915
RE244 Paris in Construction, No.2 1907
Etching 188 x 126 mm

Capt. Sir Nevile Rodwell Wilkinson, Ulster King of Arms (1869-1940)
Associate 1909
RE245 Bookplate of John George, 3rd Earl of Durham 1905
Engraving 125 x 92 mm

Henry Sheppard Dale (1852-1921)
Associate 1909
RE246 Fishmarket, Rye 1908
Etching 152 x 209 mm

Katharine Kimball (active 1906, d.1949)
Associate 1909
RE247 School of Medicine, Paris
Etching 178 x 119 mm

Alfred Edward Borthwick, R.S.A., P.R.S.W. (1871-1955)
Associate 1909
RE248 The Harbour Lantern 1909
Etching 300 x 206 mm

Frederick Marriott (1860-1941)
Associate 1909, Fellow 1924
RE249 On a Bruges Canal 1909
Etching 276 x 299 mm

William Lee-Hankey, R.W.S. (1869-1952)
Associate 1909, Fellow 1911
RE250 Waiting
Etching & aquatint 310 x 375 mm

The Hon. Walter John James, Lord Northbourne (1869-1932)
Associate 1909, Fellow 1912
RE251 Wooded Hillside
Etching 253 x 199 mm

Frank Milton Armington (1876-1941)
Associate 1910
RE252 Portal of Rathaushof, Rothenburg 1909
Etching 337 x 264 mm

Tavic Frantisek Simon (1877-1942)
Associate 1910
RE253 Bouquinistes, Paris
Etching 162 x 170 mm

Nelson Dawson, R.W.S. (1859-1941)
Associate 1910, Fellow 1915
RE254 Fishing Boat caught in a Gale
Etching 205 x 256 mm

Frederick Carter (1885-1967)
Associate 1910
RE255 L'Allegro 1909
Etching 265 x 121 mm

Janet S.C. Simpson (b.1873, still active 1938)
Associate 1910
RE256 Water Babies
Etching 193 x 350 mm

Alfred Charles Stanley Anderson, C.B.E., R.A.
(1884-1966)
Associate 1910, Fellow 1923
RE257 Somerset House 1910
Etching 218 x 388 mm

Albany E. Howarth (1872-1936)
Associate 1910
RE258 Old Houses
Etching 199 x 153 mm

James Hamilton Mackenzie, A.R.S.A., R.S.W.
(1875-1926)
Associate 1910
RE259 Italian Landscape
Etching & drypoint 224 x 200 mm

Charles Henry Baskett (1872-1953)
Associate 1911, Fellow 1918
RE260 West Mersea Marshes
Aquatint 184 x 309 mm

**Henry John (Jack) Fanshawe Badeley, Lord
Badeley, K.C.B., C.B.E.** (1874-1951)
Associate 1911, Fellow 1914
RE261 Bookplate for House of Lords' Library 1910
Engraving 93 x 120 mm

Robert Wright Stewart
Associate 1911
RE262 Dysart
Etching 149 x 253 mm

Sir Sidney Colvin (1845-1927)
Hon. R.E. 1911

Edward Fairbrother Strange, C.B.E. (1862-1929)
Hon. R.E. 1911

Myra Kathleen Hughes (1877-1918)
Associate 1911
RE263 Old Norwegian Bridge
Etching 300 x 210 mm

Alick George Horsnell (active 1909, d.1916)
Associate 1912
RE264 Palermo, Sicily 1911
Etching 150 x 200 mm

Bernard Eyre (1886-1950)
Associate 1912
RE265 On Cader Idris
Etching 149 x 223 mm

Percy Lancaster (1878-1950)
Associate 1912
RE266 The Poke Bonnet
Etching 173 x 125 mm

H. Boardman Wright (1888-1915)
Associate 1912
RE267 An Ashford Tileworks
Etching 174 x 262 mm

Frederick Charles Richards (1887-1932)
Associate 1913, Fellow 1921
RE268 St. Angelo, Perugia 1912
Etching 202 x 250 mm

William Palmer Robins, R.W.S. (1882-1959)
Associate 1913, Fellow 1917
RE269 An Old Barn 1912
Etching 110 x 162 mm

Sydney Albert Gammell (active 1909, d.1946)
Associate 1913
RE270 Peterborough
Etching 163 x 238 mm

John Robert Keitley Duff (1862-1938)
Associate 1914, Fellow 1919
RE271 Feeding the Sheep
Etching 226 x 149 mm

Ernst Viktor Hällgrën (1889-1944)
Associate 1914
RE272 Interior 1913
Etching 320 x 296 mm

Walter Monckton Keesey (1887-1970)
Associate 1914
RE273 Westminster 1913
Etching 155 x 133 mm

Raymond Ray-Jones (1886-1942)
Associate 1914, Fellow 1926
RE274 La Façade, St. Germain l'Auxerrois, Paris
Etching 214 x 214 mm

Gurnell C. Jennis (fl.1910-1936)
Associate 1914
RE275 On Strike
Etching 175 x 98 mm

William Walker (1878-1961)
Associate 1914
RE276 Middleburg, Holland
Etching 278 x 189 mm

Niels Moller Lund (1863–1916)
Associate 1915
RE277 Pevensey
Etching 257 x 359 mm

Frederick H. Townsend (1868–1920)
Associate 1915
RE278 Patience 1914
Etching 179 x 232 mm

Auguste Lepère (1849–1918)
Hon. R.E. 1914

Sidney Tushingham (1884–1968)
Associate 1915
RE279 Portrait of my Wife
Etching 276 x 127 mm

Julia E. Clutterbuck (active 1909, d.1932)
Associate 1916
RE280 Tide Mill, Fishbourne
Etching 150 x 100 mm

Frederick Landseer Maur Griggs, R.A. (1876–1938)
Associate 1916, Fellow 1918
RE281 Priory Farm 1913
Etching 116 x 152 mm

Arthur James Turrell (b.1871, still active 1936)
Associate 1916
RE282 Portrait, after Moreelse 1907
Etching 428 x 322 mm

Herbert Gordon Warlow (1885–1942)
Associate 1916
RE283 South Porch, King's College Chapel,
Cambridge
Etching 367 x 297 mm

Henry Winslow
Associate 1916

Dorothy G. Woollard (1886–1986)
Associate 1916, Fellow 1925
RE284 Burnham Beeches 1915
Etching 251 x 225 mm

Margaret Stirling Dobson (fl.1905–1936)
Associate 1917
RE285 The West Wind
Drypoint 350 x 300 mm

Laura Sylvia Gosse (1881–1968)
Associate 1917, Fellow 1936
RE286 Costers by Sink
Etching 234 x 152 mm

Charles Haigh Wood (1856–1927)
Associate 1917
RE287 Feeding Calves
Etching 201 x 227 mm

William Westley Manning (1868–1954)
Associate 1917
RE288 Valley of the Adur, Sussex
Etching 175 x 253 mm

Leonard Russell Squirrell, R.W.S. (1893–1979)
Associate 1917, Fellow 1919
RE289 At Campsea Ash, Suffolk 1914
Etching 183 x 248 mm

Henry Percy Huggill (1886–1957)
Associate 1917
RE290 Seville, Spain
Etching 216 x 266 mm

William Walcot (1874–1943)
Associate 1918, Fellow 1920
RE291 Tragedy of Sophocles
RE292 Marble Arch
Etchings 440 x 465; 128 x 216 mm

Reginald Herbert Green (1884–1971)
Associate 1918
RE293 The Two Windmills
Etching 187 x 225 mm

George Marples (1869–1939)
Associate 1918
RE294 Pan
Etching 501 x 375 mm

Doris Boulton (1892–1961)
Associate 1918
RE295 Study of a Head
Engraving 284 x 170 mm

George Soper (1870–1942)
Associate 1918, Fellow 1920
RE296 A Devon Shoeing-smith
Etching 176 x 125 mm

Leslie Moffat Ward (1888–1978)
Associate 1918, Fellow 1936
RE297 Tower of Westminster, from Vauxhall
Bridge 1917
Etching & drypoint 199 x 174 mm

Greta Delleany (active 1913, d.1968)
Associate 1919
RE298 Animal Studies
Etching 250 x 176 mm

Sir Aston Webb, K.C.V.O., C.B., P.R.A. (1849–1930)
Hon. R.E. 1919

Campbell Dodgson, C.B.E. (1867–1948)
Hon. R.E. 1919

Alexander Théophile Steinlen (1859–1923)
Hon. R.E. 1919

Edmund Blampied (1886-1966)
Associate 1920, Fellow 1929
RE299 Flies
Drypoint 126 x 175 mm

Hester Frood (1882-1971)
Associate 1920
RE300 Apse of Avila Cathedral
Etching 155 x 161 mm

Noel Rooke (1881-1953)
Associate 1920
RE301 The Edge of the Wood 1920
Woodcut 299 x 491 mm

Gwendolen Raverat (1885-1957)
Associate 1920, Fellow 1934
RE302 The Gooseherd 1919
Wood engraving 101 x 191 mm

Katharine Cameron, R.S.W. (1874-1965)
Associate 1920, Fellow 1964
RE303 Grass of Parnassus
Etching 364 x 136 mm

Ernest Herbert Whydale (1886-1952)
Associate 1920
RE304 Nearing the Hill-top
Etching 220 x 263 mm

Nora Molly Campbell (1888-1971)
Associate 1920
RE305 Revelry 1915
Etching 115 x 180 mm

Rowland Roy Gill (active 1910, d.1965)
Associate 1920
RE306 The Roman Road
Etching 174 x 244 mm

Théodore Roussel (1847-1926)
Associate 1921
RE307 Bathers
Etching 51 x 156 mm

Gerald Leslie Brockhurst, R.A. (1890-1978)
Associate 1921, Fellow 1921
RE308 By the Mirror
Etching 146 x 101 mm

Sydney Long (1872-1955)
Associate 1921
RE309 On the Wandle
Etching 295 x 259 mm

Lilian Whitehead (b.1894, still active 1934)
Associate 1921
RE310 Sewing 1920
Etching 125 x 87 mm

Sir Henry George Rushbury, K.C.V.O., C.B.E., R.A., R.W.S. (1889-1968)
Associate 1921, Fellow 1922
RE311 The Fish Market, Marseilles
Etching & drypoint 251 x 174 mm

John Laviers Wheatley, A.R.A., R.W.S. (1892-1955)
Associate 1921

Robert Sargent Austin, P.R.E., R.A., P.R.W.S. (1895-1973)
Associate 1921, Fellow 1928, President 1962-70
RE312 The Trace Horse 1921
Etching 167 x 202 mm

Robert Charles Peter (1888-1980)
Associate 1921, Fellow 1973
RE313 Dawn
Mezzotint 306 x 387 mm

Malcolm Charles Salaman (1855-1940)
Hon. R.E. 1921

Norman Janes, R.W.S. (1892-1980)
Associate 1921, Fellow 1938
RE314 An Old Man 1921
Etching 162 x 203 mm

Claude Allin Shepperson, A.R.A., A.R.W.S. (1867-1921)
Associate 1921

John Frederick Greenwood (1885-1954)
Associate 1922, Fellow 1939
RE315 The Tramp
Wood engraving 126 x 100 mm

Charles William Taylor (1878-1960)
Associate 1922, Fellow 1930
RE316 Leigh, Essex
Wood engraving 109 x 154 mm

John Charles Moody (1884-1962)
Associate 1922, Fellow 1946
RE317 The Brass Market 1921
Etching 300 x 224 mm

Solomon van Abbé (1883-1955)
Associate 1923
RE318 Study of the Nude
Drypoint 240 x 210 mm

John Nicolson, R.S.W. (1891-1951)
Associate 1923
RE319 Calf Drinking
Etching 210 x 263 mm

Arthur Ralph Middleton Todd, R.A., R.W.S. (1891-1966)
Associate 1923, Fellow 1930

RE320 The Old Coster 1922
Etching 176 x 126 mm

Iain Macnab of Barachastlain (1890-1968)
Associate 1923, Fellow 1935
RE321 The Palace Guard
Aquatint 150 x 200 mm

Jan M. Daum (fl.1915-1929)
Associate 1923
RE322 The Morning Rasher
Drypoint 200 x 160 mm

Rosa Somerville Hope (1902-1972)
Associate 1923
RE323 Man Playing Guitar
Etching 215 x 175 mm

Job Nixon, R.W.S. (1891-1938)
Associate 1923, Fellow 1933
RE324 Anticoli
Drypoint 325 x 385 mm

Wendela Boreel (1895-1985)
Associate 1923
RE325 Study of a Baby
Etching 218 x 156 mm

Dame Laura Knight, D.B.E., R.A., R.W.S.
(1877-1970)
Associate 1924, Fellow 1932
RE326 Putting on Tights 1926
Etching 202 x 176 mm

John Macdonald Aiken, R.S.A. (1880-1963)
Associate 1924
RE327 An Old Fisherman
Drypoint 199 x 147 mm

Alec Buckels (active 1923, d.1972)
Associate 1924
RE328 Friendless 1923
Wood engraving 102 x 101 mm

Eliab George Earthrowl (b.1878, still active 1929)
Associate 1924
RE329 Bourg L'Oisans, French Alps, No.2 1923
Etching 346 x 205 mm

John Gerald Platt (1892-1975)
Associate 1924
RE330 The Jewish Rabbi
Mezzotint 268 x 226 mm

Ernest Heber Thompson (1891-1971)
Associate 1924, Fellow 1930
RE331 The Bathers 1923
Etching 198 x 250 mm

Edmund Joseph Sullivan, A.R.W.S. (1869-1933)
Associate 1924, Fellow 1931

RE332 The Rogues' March, "Don Quixote"
Etching 150 x 196 mm

Sir Frank Dicksee, K.C.V.O., P.R.A. (1853-1928)
Hon. R.E. 1925

Graham Vivian Sutherland, O.M. (1903-1980)
Associate 1925
RE333 No.49 1924
Etching 177 x 252 mm

William Martin Larkins (1901-1974)
Associate 1925
RE334 Le Quai du Rosaire, Bruges
Etching 200 x 252 mm

Geoffrey Heath Wedgwood (1900-1977)
Associate 1925, Fellow 1934
RE335 The Empire 1924
Engraving 158 x 296 mm

Ian Strang (1886-1952)
Associate 1926, Fellow 1930
RE336 San Gil, Burgos 1924
Etching 225 x 276 mm

Paul Dalou Drury, P.R.E. (1903-1987)
Associate 1926, Fellow 1932, President 1970-75
RE337 Negro 1925
Etching 123 x 100 mm

Louis Conrad Rosenberg (1890-1983)
Associate 1926, Fellow 1931
RE338 St. Etienne, Toulouse
Etching 253 x 117 mm

Charles Sidney Cheston, R.W.S. (1882-1960)
Associate 1927
RE339 Landscape with Farm
Etching 175 x 326 mm

Arthur Joseph Gaskin (1862-1928)
Associate 1927
RE340 Gathering Winter Fuel
Wood engraving 145 x 89 mm

James Ardern Grant (1887-1973)
Associate 1928, Fellow 1932
RE341 Zeta 1929
Drypoint 210 x 136 mm

Harry Morley, A.R.A., R.W.S. (1881-1943)
Associate 1929, Fellow 1931
RE342 Hylas and the Nymphs
Engraving 176 x 125 mm

Charles Frederick Tunnicliffe, R.A. (1901-1979)
Associate 1929, Fellow 1934
RE343 The Bull 1926
Etching 226 x 277 mm

Evelyn Gibbs (1905-1991)
Associate 1929, Fellow 1973
RE344 Portrait of a Girl 1928
Etching 180 x 152 mm

Sir William Llewellyn, P.R.A. (1858-1941)
Hon. R.E. 1929

Arthur John Trevor Briscoe (1873-1943)
Associate 1930, Fellow 1933
RE345 James and John 1928
Etching 251 x 360 mm

Frederick Austin (1902-1990)
Associate 1930, Fellow 1951
RE346 Flight into Egypt 1927
Drypoint 174 x 209 mm

Clare Veronica Hope Leighton (1898-1989)
Associate 1930, Fellow 1934
RE347 Brother Juniper
Wood engraving 119 x 77 mm

Joseph Webb (1908-1962)
Associate 1930
RE348 The Rat Barn 1929
Etching 175 x 300 mm

Stanley Roy Badmin, R.W.S. (1906-1989)
Associate 1931, Fellow 1935
RE349 Burford, Oxfordshire 1930
Etching 127 x 185 mm

James Bateman, R.A., A.R.W.S. (1893-1959)
Associate 1931
RE350 Deer in Richmond Park 1930
Wood engraving 147 x 180 mm

Eric Fitch Daglish (1894-1966)
Associate 1931

Sir William Russell Flint, R.A., P.R.W.S. (1880-1969)
Associate 1931, Fellow 1933
RE351 Spanish Wheelwrights
Drypoint 234 x 315 mm

Stephen Frederick Gooden, R.A. (1892-1955)
Associate 1931, Fellow 1933
RE352 The Satyrs 1937
Engraving diameter 150 mm

Norman Hirst (1862-1956)
Associate 1931
RE353 Rye Mill 1926
Mezzotint 223 x 293 mm

Julius Komjati (1894-1958)
Associate 1931
RE354 The Prayer 1928
Etching 252 x 203 mm

William Washington (1885-1956)
Associate 1931
RE355 St. Olave's, Southwark 1929
Engraving 348 x 302 mm

James McIntosh Patrick, R.S.A. (1907-1998)
Associate 1932
RE356 Les Ramparts, Les Baux 1927
Etching 175 x 250 mm

Leonard Griffith Brammer (1906-1994)
Associate 1932, Fellow 1956
RE357 Burslem 1930
Etching 243 x 360 mm

Laurence Binyon, C.H. (1869-1943)
Hon. R.E. 1932

Arthur Mayger Hind, O.B.E. (1880-1957)
Hon. R.E. 1932

Wilfred Fairclough, R.W.S. (1907-1996)
Associate 1934, Fellow 1946
RE358 The Arrival 1933
Etching 252 x 162 mm

Henry Martyn Lack (1909-1979)
Associate 1934, Fellow 1948
RE359 Winter Landscape, Northants 1931
Etching 92 x 171 mm

John Taylor Arms (1887-1953)
Associate 1934
RE360 Shadows of Venice 1930
Etching 256 x 302 mm

Arthur Stewart Hunt Mills (1897-1968)
Associate 1935, Fellow 1948
RE361 Gadshill 1933
Wood engraving 80 x 140 mm

Raymond Teague Cowern, R.A., R.W.S. (1913-1986)
Associate 1935, Fellow 1946
RE362 Suspense 1934
Etching 180 x 170 mm

Robert Stanley Gorrell Dent, R.W.S. (1909-1991)
Associate 1935, Fellow 1946
RE363 Visiting Day 1934
Etching 158 x 202 mm

Daphne Lindner (b.1912, still active 1937)
Associate 1935
RE364 Convalescence
Etching 181 x 175 mm

Charles H. Spencer (1909-1982)
Associate 1935
RE365 Farmstead, Fragnito Monforte 1934
Etching 147 x 213 mm

Eric Wilfred Taylor (b.1909)
Associate 1935, Fellow 1948
RE366 Patience 1934
Etching 164 x 151 mm

Nora Spicer Unwin (1907-1982)
Associate 1935, Fellow 1946
RE367 Teazle 1931
Wood engraving 177 x 127 mm

Lawrence Leon Louis Josset (1910-1995)
Associate 1936, Fellow 1951
RE368 The Coat of Many Colours 1936
Etching 157 x 205 mm

Barbara Moray Williams (1911-1975)
Associate 1936, Fellow 1973
RE369 Danse Macabre 1932
Wood engraving 141 x 122 mm

Eveleen Buckton (1872-1962)
Associate 1937
RE370 Cley, Norfolk
Etching 102 x 248 mm

John Farleigh, C.B.E. (1900-1965)
Associate 1937, Fellow 1948
RE371 Milk Thistle 1936
Wood engraving 171 x 107 mm

Harold Wilfred Sayer (1913-1993)
Associate 1937, Fellow 1982
RE372 The Hostel 1936
Etching 214 x 156 mm

Robin Tanner (1904-1988)
Associate 1937
RE373 Hedge Flowers 1936
Etching 227 x 157 mm

James Laver, C.B.E., R.S.A. (1899-1975)
Hon. R.E. 1937

Hubert Andrew Freeth, R.A., R.W.S. (1912-1986)
Associate 1937, Fellow 1946
RE374 Black Country 1935
Etching 124 x 175 mm

Eleanor Erlund Hudson, R.W.S. (b.1912)
Associate 1937, Fellow 1946
RE375 Lamplight
Etching 174 x 194 mm

Ruth Spencer Aspden (b.1909)
Associate 1937
RE376 The Bedroom
Etching 194 x 206 mm

Clifford Webb (1895-1972)
Associate 1937, Fellow 1948

RE377 Wood Farm
Wood engraving 189 x 228 mm

Esmé Mary Evelyn Currey (c.1905-1973)
Associate 1937
RE378 Evening, Alfriston
Etching 113 x 123 mm

George Buday (1907-1990)
Associate 1938, Fellow 1953
RE379/1 Mary Magdalen
RE379/2 The Annunciation
Wood engravings 99 x 58; 100 x 59 mm

Vyvyan Grindley-Ferris (active 1935, d.1941)
Associate 1938
RE380 Earl's Court
Etching & aquatint 192 x 394 mm

Frank Joseph Archer, R.W.S. (1912-1995)
Associate 1940, Fellow 1960
RE381 Peasants Bathing 1939
Etching 188 x 169 mm

Barbara Greg (1900-1983)
Associate 1940, Fellow 1946
RE382 Edge of the Wood
Wood engraving 182 x 152 mm

Joan Hassall (1906-1988)
Associate 1940, Fellow 1948
RE383 The Water Splash 1937
Wood engraving 76 x 108 mm

Harold H. Shelton (1913-1994)
Associate 1940
RE384 Gathering Faggots
Etching 178 x 243 mm

Nandor Lajos Varga (1895-1978)
Associate 1940, Hon. R.E. 1970
RE385 Head of Girl 1927
Drypoint 164 x 171 mm

Agnes Miller Parker (1895-1980)
Associate 1940, Fellow 1953
RE386 The Challenge 1934
Wood engraving 142 x 163 mm

Denise Jeanne Marie Lebreton Brown
(1911-1998)
Associate 1941, Fellow 1959
RE387 L'Étudiant
Drypoint 189 x 172 mm

Marion Rhodes (1907-1998)
Associate 1941, Fellow 1953
RE388 King's Cross Station
Etching 159 x 206 mm

Alan Sugden
Associate 1941
RE389 A Journey
Etching 154 x 177 mm

Sara Sproule, R.W.S. (b.1914)
Associate 1942, Fellow 1948
RE390 Barber's Shop, Ambleside 1941
Etching 224 x 202 mm

Robert Ball (b.1918)
Associate 1943, Fellow 1978
RE391 Self Portrait 1937
Etching 187 x 133 mm

Gwendolen Marie May (1903-1992)
Associate 1943, Fellow 1966
RE392 The Designers' Room
Etching 263 x 212 mm

Harold Thornton (1892-1958)
Associate 1943
RE393 Widdecombe-in-the-Moor
Etching 208 x 324 mm

David William Brokman Davis (b.1892)
Associate 1944
RE394 Flower Cluster 1933
Etching 100 x 99 mm

Gwyneth Joan Eedy
Associate 1944
RE395 Europa
Wood engraving 64 x 89 mm

Hazel W. Harrison (b.1918)
Associate 1944
RE396 Mattress Fillers
Etching 140 x 153 mm

Edgar Owen Jennings, R.W.S. (1899-1985)
Associate 1944, Fellow 1970
RE397 Sand Cart, Siena
Engraving 250 x 303 mm

James Thomas Armour Osborne (1907-1979)
Associate 1944, Fellow 1956
RE398/1 Winter Sleet 1943
Wood engraving 184 x 235 mm
RE398/2 The Rare Bittern of the Fens
Woodcut 710 x 457 mm

Alan Herbert Carr Linford (b.1926)
Associate 1946
RE399 View at Clappersgate, Westmorland 1945
Etching 169 x 146 mm

Jean Graham Harper, R.W.S. (b.1921)
Associate 1946, Fellow 1956

RE400 The New Dress 1945
Etching 158 x 136 mm

Russell Reeve (1895-1970)
Associate 1946, Fellow 1956
RE401 In the Cow Yard
Etching 190 x 239 mm

Henry W. Wilkinson (b.1921)
Associate 1946, Fellow 1970
RE402 The Belfry
Etching 186 x 140 mm

James Edward Bostock (b.1917)
Associate 1947, Fellow 1961
RE403 Ducks in the Wood 1946
Wood engraving 126 x 175 mm

G. W. Lennox Paterson (1915-1986)
Associate 1947, Fellow 1973
RE404 Christmas Decoration 1946
Wood engraving 117 x 101 mm

John S. O'Connor, R.W.S. (b.1913)
Associate 1947
RE405 Little Garden in the Evening 1946
Wood engraving 128 x 174 mm

Harold James Lean Wright (d.1961)
Hon. R.E. 1946

Murray Macpherson Tod (1909-1974)
Associate 1947, Fellow 1953
RE406 Winter 1941
Etching 123 x 199 mm

Anthony Gross, C.B.E., R.A., A.R.W.S. (1905-1984)
Associate 1947, Hon. R.E. 1979
RE407 Girl on Bicycle, Fulham 1947
Etching 170 x 200 mm

Fred Middlehurst (1918-1982)
Associate 1945
RE408 Jean
Mezzotint 212 x 159 mm

Emerson Harold Groom (1891-1983)
Associate 1945

Robert John Gibbings (1889-1958)
Associate 1948

John Copley (1875-1950)
Associate 1948, Fellow 1948

Cyril Edward Deakins (b.1916)
Associate 1948
RE412 Theatre Scene 1939
Etching 148 x 199 mm

Nigel Lambourne (1919-1988)
Associate 1948

RE413 On the Beach 1947
Etching 147 x 165 mm

Maud Sharp (1897-1963)
Associate 1948
RE414 Wells Cathedral 1946
Etching 200 x 258 mm

Erik Smith, R.W.S. (1914-1973)
Associate 1948, Fellow 1959
RE415 Dumfries Churchyard 1947
Etching 182 x 246 mm

Geoffrey Wales (1912-1990)
Associate 1948, Fellow 1961
RE416 The Bay
Wood engraving 150 x 180 mm

Sir Muirhead Bone (1876-1953)
Hon. R.E. 1948

Gertrude Hermes, R.A. (1901-1983)
Associate 1948, Fellow 1951, Hon. R.E. 1967
RE420/1 The Prawn 1950
RE420/2 Self Portrait No.19 1949
Wood engravings & lino block 176 x 249; 204 x 153 mm

James Taylor Dolby (1909-1975)
Associate 1949, Fellow 1967
RE417 Minesweeping Gear 1947
Wood engraving 129 x 153 mm

**Harry Norman Eccleston, O.B.E., P.P.R.E.,
R.W.S.** (b.1923)
Associate 1949, Fellow 1961, President 1975-89
RE418 The Cabbage Patch 1948
Etching 128 x 160 mm

Margery Gill (d.1983)
Associate 1949
RE419 Girl with Beads 1947
Etching 156 x 105 mm

Arthur Hackney, R.W.S. (1925-1994)
Associate 1949, Fellow 1960
RE421 Broken Lamp
Engraving 131 x 108 mm

Anne Scott
Associate 1949
RE422 Snow-scene 1947
Etching & aquatint 253 x 221 mm

Jozef Sekalski (1904-1972)
Associate 1949
RE423 Untitled 1945
Wood engraving 176 x 134 mm

Prudence Eaton Seward (b.1926)
Associate 1949, Fellow 1953
RE424 The Model 1948
Etching 189 x 136 mm

Thomas William Ward, R.W.S. (b.1918)
Associate 1949, Fellow 1953
RE425 Hammersmith Turnout 1948
Engraving 140 x 216 mm

Harold Charles Bartlett, P.P.R.W.S. (b.1921)
Associate 1950, Fellow 1961,
RE426 Girl putting on Shoes
Etching 202 x 154 mm

Charles Herbert Arthur Chaplin (1907-1986/7)
Associate 1950, Fellow 1961
RE427 Stoat 1950
Engraving 92 x 201 mm

George Edward Mackley (1900-1983)
Associate 1950, Fellow 1961
RE428 By the Great Ouse
Wood engraving 77 x 103 mm

Owen Olley
Associate 1950
RE429 Monmouth Street Café
Etching 192 x 236 mm

Winifred J. Taylor (b.1924)
Associate 1950
RE430 Europa 1947
Etching 245 x 206 mm

David Vaughan Wicks (1918-1996)
Associate 1950, Fellow 1961
RE431 Trees 1949
Etching 138 x 201 mm

Brian Bradshaw (b.1923)
Associate 1951
RE432 The Chinese Bed
Etching & aquatint 176 x 148 mm

Alfred Hackney, R.W.S. (b.1926)
Associate 1951
RE434 Puppet Maker 1950
Etching & aquatint 226 x 325 mm

Philip O. Jennings (1921-1983)
Associate 1951, Fellow 1978
RE435 Piccadilly Toy Seller 1950
Etching & aquatint 211 x 170 mm

Kenneth Herbert Oliver, R.W.S. (b.1923)
Associate 1951, Fellow 1978
RE436 Self Portrait
Etching & aquatint 197 x 151 mm

Thomas Donald Plenderleith (1921-1995)
Associate 1951, Fellow 1961
RE437 War Dance 1951
Wood engraving 155 x 230 mm

John Vivian Roberts, R.W.S. (b.1923)
Associate 1951, Fellow 1965
RE438 Interval 1950
Etching & aquatint 174 x 151 mm

David Smith (b.1920)
Associate 1951, Fellow 1973
RE439 Temple
Etching & aquatint 199 x 249 mm

Norman Webster, R.W.S. (b.1924)
Associate 1951, Fellow 1973
RE440 A Saddle-billed Stork, Tropical Africa
Etching 240 x 163 mm

John Gilbey Bowles (b.1929)
Associate 1952
RE441 Richmond 1951
Etching & aquatint 318 x 354 mm

Gilbert Henry Mason (1913-1972)
Associate 1952
RE442 The Birdcage
Drypoint 222 x 175 mm

James William Lovegrove (1922-1997)
Associate 1952
RE443 The Queen's Head
Etching 160 x 250 mm

Bernard Brett
Associate 1952
RE444 Saloon Bar 1951
Etching 170 x 173 mm

Peter Bottomley (1927-1982)
Associate 1952
RE445 Repairs 1951
Aquatint 228 x 269 mm

Gwenda Morgan (1908-1991)
Associate 1952, Fellow 1973
RE446 Flowers in Little Jug
Wood engraving 127 x 123 mm

Christopher John Alexander (1926-1982)
Associate 1952
RE447 The Bright Hour 1951
Etching 225 x 144 mm

Ian Churston Barnard (b.1927)
Associate 1953
RE448 Northern Landscape 1952
Etching 131 x 209 mm

Peter Arthur Downing
Associate 1953
RE449 Fishermen
Etching & aquatint 138 x 170 mm

Edward Carlos Francis
Associate 1953
RE450 The Orchard
Etching & aquatint 215 x 209 mm

George Douglas Halliday
Associate 1953
RE451 Touch 1951
Etching 156 x 224 mm

Keith Mackenzie (1924-1993)
Associate 1953
RE452 68 Kensington Church Street 1952
Etching 283 x 148 mm

Francis John Winter (1901-1996)
Associate 1953, Fellow 1973
RE453 Succulents 1950
Wood engraving 253 x 202 mm

James K. Young (b.1924)
Associate 1953, Fellow 1983
RE454 Crowd at Tivoli
Etching 182 x 241 mm

John Buckland Wright (1897-1954)
Associate 1953
RE455 Camber Sands 1953
Wood engraving 176 x 253 mm

Walter Hudspith
Associate 1954

Philip Thomas Langford Reeves, R.S.A. (b.1931)
Associate 1954, Fellow 1965
RE457/1 A Cotswold Hill
Aquatint 495 x 824 mm
RE457/2 London Bridge Underground
Etching 120 x 154 mm

Gavin R. Rowe
Associate 1954
RE458 Seated Figure 1952
Etching 146 x 117 mm

Ronald William Maris (b.1932)
Associate 1955
RE459 The Flies
Etching & drypoint 272 x 408

John Wright (b.1927)
Associate 1955, Fellow 1975
RE460 The Gatherer 1953
Etching 250 x 295 mm

Thelma Muriel Blakely (1927-1978)
Associate 1955, Fellow 1966
RE461 Flora
Wood engraving 178 x 127 mm

Henry James Starling (1905-1996)
Associate 1955
RE462 Wymondham Abbey
Etching 220 x 298 mm

Maurice Heap (b.1930)
Associate 1955
RE463 Corner Restaurant
Etching & aquatint 162 x 200 mm

**Prof. Sir Albert Richardson, K.C.V.O., P.R.A.,
Hon. R.W.S.** (1880-1964)
Hon. R.E. 1955

Bryan Stephen Hornsby (b.1919)
Associate 1955
RE464 Girl in Taxi
Etching 163 x 255 mm

Anthony Lewis Pope
Associate 1955
RE465 English Sundew
Etching & aquatint 148 x 97 mm

Margaret Butler (b.1932)
Associate 1955
RE466 John William Street 1953
Etching 188 x 338 mm

Frank Vernon Martin (b.1921)
Associate 1955, Fellow 1961
RE467 A Mexican Journey
Wood engraving 140 x 96 mm

Erich Wolsfeld (1884-1956)
Associate 1955
RE468 Arab Shepherds
Etching 302 x 376 mm

Nevil Shaw, R.W.S. (1915-1989)
Associate 1955
RE469 Architectural Composition No.1 1955
Etching 250 x 409 mm

V.S. Adurker
Associate 1956

Patrick Burke (b.1932)
Associate 1956
RE470 Loading-up 1956
Etching & aquatint 224 x 100 mm

Brian Perrin, A.R.W.S. (b.1932)
Associate 1956-65, re-elected Fellow 1988
RE471 Untitled, urban scene 1955
Etching & drypoint 439 x 340 mm

Richard Fozard (b.1925)
Associate 1956

RE472 Piazza Anticoli 1956
Etching & aquatint 274 x 343 mm

Dolf Rieser (1898-1983)
Associate 1956, Fellow 1968
RE473 Fishing Birds
Aquatint 300 x 424 mm

Arthur Henderson Hall (1906-1983)
Associate 1956, Fellow 1961
RE474 The Lobster Pot 1957
Etching 120 x 197 mm

Geoffrey Sparrow (d.1970)
Associate 1956
RE475 "Who-oop!"
Aquatint 147 x 195 mm

David Jupp (b.1929)
Associate 1956

Albert Phillip (b.1915)
Associate 1956
RE477 A Meal
Etching 172 x 162 mm

Edward Croft Murray (1907-1980)
Hon. R.E. 1956

**Lynd Ward, Pres. Soc. American Graphic
Artists**
Hon. R.E. 1956

Michael John Blaker (b.1928)
Associate 1957, Fellow 1975
RE478 Portrait of a Girl 1956
Drypoint 200 x 148 mm

John Petts (1914-1991)
Associate 1957
RE479 The Three Crags
Wood engraving 102 x 128 mm

**Sir Charles Wheeler, K.C.V.O., C.B.E., P.R.A.,
Hon. R.W.S.** (1892-1974)
Hon. R.E. 1957

David Anthony Cox
Associate 1958
RE481 Foundry Workers 1957
Etching & drypoint 147 x 263 mm

Walden L. Tucker
Associate 1958
RE482 Richmond Terrace
Etching 122 x 251 mm

Derrick Harris (1919-1960)
Associate 1958
RE483 Book Lovers
Wood engraving 209 x 194 mm

Alison Mary Prince (b.1931)
Associate 1958
RE484 The Bull 1957
Etching & aquatint 201 x 279 mm

David Euan Jennings (b.1933)
Associate 1958, Fellow 1966
RE485 Donkeys 1957
Linocut 355 x 493 mm

Newton Taylor (b.1911)
Associate 1958
RE486 Open-cast Mining at Wentworth, Yorks 1957
Etching & aquatint 200 x 250 mm

Prof. Peter A. Green, O.B.E. (b.1934)
Associate 1958, Fellow 1978
RE487 Sea Wall and Anchors 1957
Linocut 305 x 409 mm

Derek Sidney Collard
Associate 1959
RE488 Dartmouth Gardens 1959
Etching & aquatint 251 x 315 mm

Marcia Lane Foster (1897-1983)
Associate 1959
RE489 Sunbathers
Wood engraving 146 x 216 mm

Kenneth Arthur Lindley (1928-1986)
Associate 1959, Fellow 1975
RE490 Adam's Grave, Alton Priory
Wood engraving 69 x 178 mm

Dr. Jennifer Joan Dickson, R.A. (b.1936)
Associate 1959, Fellow 1965
RE491 Girl doing her Hair
Etching & aquatint 178 x 126 mm

Keith R. Armour (b.1931)
Associate 1960
RE492 Hill Town 1958
Etching & aquatint 280 x 185 mm

John Anthony Brown (b.1932)
Associate 1960
RE493 St. Francis 1959
Wood engraving 101 x 228 mm

John Boys Drawbridge, M.B.E.
Associate 1960
RE494 The Loire Valley near Saumur 1958
Etching & aquatint 262 x 481 mm

Raymond Humphrey Millis Hawthorn (1917-1997)
Associate 1960, Fellow 1975
RE495 The Struggle for Greece
Wood engraving 176 x 105 mm

Rachel Ann Le Bas (b.1923)
Associate 1960, Fellow 1969
RE496 Washerwomen by Lake Garda
Etching & aquatint 253 x 302 mm

Eric Ramsden (1927-1997)
Associate 1960
RE497 Trees at Laverstoke
Wood engraving 66 x 103 mm

Rachel Roberts (b.1908)
Associate 1960
RE498 February's Flowers
Wood engraving 196 x 180 mm

Jennifer N. Black
Associate 1961
RE499 Chiswick Park 1959
Linocut 254 x 703 mm

Leslie Jones (b.1934)
Associate 1961
RE500 Crocifissione 1959
Etching & aquatint 268 x 336 mm

Mary Malenoir (b.1940)
Associate 1961, Fellow 1984
RE501 Bean Pickers 1960
Etching 134 x 146 mm

James Morrison Townley (1917-1995)
Associate 1961
RE502 Bridge at Hadley Wood 1960
Intaglio 302 x 455 mm

Sally Frances McLaren (b.1936)
Associate 1962, Fellow 1973
RE503 Celebration 1961
Intaglio 610 x 500 mm

Barbara Newcomb (b.1936)
Associate 1962-68, re-elected Fellow 1984
RE504 Thames [1] 1961
Aquatint 315 x 500 mm

Thomas Walsh (b.1938)
Associate 1962, Fellow 1975
RE505 Port Dundas, Glasgow
Aquatint 301 x 400 mm

Margaret Payne (b.1937)
Associate 1962, Fellow 1975
RE506 Cornish Harbour
Etching & aquatint 345 x 378 mm

George W. Tute (b.1933)
Associate 1962, Fellow 1973
RE507 Dandelions   1960
Wood engraving 105 x 124 mm

Michael T.J. Reynolds
Associate 1962
RE508 Wedding   1961
Etching 606 x 249 mm

Garrick Salisbury Palmer (b.1933)
Associate 1963, Fellow 1970
RE509 Landscape   1963
Wood engraving 124 x 190 mm

Vernon Mills (b.1931)
Associate 1963
RE510 Hog Weed & Willows   1959
Wood engraving 145 x 98 mm

Victor Gyōzō Ambrus (b.1935)
Associate 1964, Fellow 1973
RE511 Daphnis and Chloe   1959
Etching 283 x 156 mm

Michael Fairclough (formerly Cryer) (b.1940)
Associate 1964, Fellow 1973
RE512 Sea-Wall   1963
Etching & aquatint 211 x 423 mm

David Robert Sawyers (b.1941)
Associate 1964
RE513 Lime Works   1964
Etching 356 x 440 mm

Sylvia Melland (1906-1993)
Associate 1965, Fellow 1973
RE514/1 Icarus   1964
RE514/2 Oyster Catcher   1964
Etchings & aquatint 342 x 496; 295 x 499 mm

Herbert John Jackson (b.1938)
Associate 1965, Fellow 1975
RE515 Fish Wharf/1   1966
Linocut 390 x 522 mm

Trevor Abbott Allen (b.1939)
Associate 1965, Fellow 1995
RE516 Tractor and Fields   1964
Silkscreen 550 x 460 mm

Jack Robert Shirreff (b.1943)
Associate 1965
RE517 Shattered Images   1965
Etching & aquatint 504 x 348 mm

Brian Collier (b.1945)
Associate 1966
RE518 Between Heaven and Earth   1965
Etching & aquatint 594 x 478 mm

Robert Tavener (b.1920)
Associate 1966, Fellow 1973
RE519 Pollarded Trees No.1   1964
Linocut 432 x 586 mm

Stanley William Hayter, C.B.E., Hon.R.A.
(1901-1988)
Hon. R.E. 1966
RE520 Lunar Rhythm   1967
Etching & aquatint 347 x 483 mm

James Richard Tierney (b.1945)
Associate 1967
RE521 Bird over Landscape   1966
Etching & aquatint 782 x 565 mm

David Shaw Barker (b.1945)
Associate 1967, Hon. R.E. 1996
RE522 Basic Vocabulary "Tea Pot"   1994
Screenprint 665 x 423 mm

Monica Poole (b.1921)
Associate 1967, Fellow 1975
RE523 Shell   1967
Wood engraving 230 x 168 mm

Robert Alan Lewis Burnand (b.1929)
Associate 1968
RE524 Fishing
Wood engraving 69 x 102 mm

Michael James Chaplin, R.W.S. (b.1943)
Associate 1968, Fellow 1973
RE525 Untitled, abstract   1967
Etching & aquatint 451 x 603 mm

Richard Jack Harley (b.1945)
Associate 1968
RE525A Stanford
Etching 308 x 435 mm

Valerie Thornton (1931-1991)
Associate 1968, Fellow 1970
RE526 Montacute House   1957
Etching & aquatint 286 x 518 mm

Sonnylal Rambissoon (1926-1995)
Associate 1968
RE527 Caribbean Heat   1964
Aquatint 329 x 500 mm

Ann Brunskill (b.1923)
Associate 1969, Fellow 1973

Peter Reddick (b.1924)
Associate 1969, Fellow 1973
RE529 Crotchet Castle   1962
Woodcut 161 x 89 mm

Eva H. Stockhaus (b.1919)
Associate 1969, Fellow 1975
RE529A Gulls in the Field 1968
Wood engraving 249 x 188 mm

John Urban (1941-1988)
Associate 1969, Fellow 1978
RE530 Tivoli 1968
Etching 310 x 472 mm

John Heagan Eames (b.1900)
Associate 1969, Fellow 1978
RE531 San Miniato al Monte, Firenze
Mezzotint 179 x 239 mm

Edward Osmond Zoltan Wade (b.1940)
Associate 1970
RE532 Untitled animal study 1969
Wood engraving 64 x 90 mm

Tranquillo Marangoni (1912-1992)
Hon. R.E. 1971
RE533 Vimini sulla spiaggia a Scheveningen 1955
Wood engraving 315 x 230 mm

Sir Walter Thomas Monnington, K.C.V.O., P.R.A., Hon. R.W.S. (1902-1976)
Hon. R.E. 1971

Joan Williams, R.W.S. (b.1922)
Associate 1971, Fellow 1975
RE534 Two Volcanos 1976
Etching & aquatint 485 x 507 mm

Jack Coutu (b.1924)
Associate 1972

Mary Louise Coulouris (b.1939)
Associate 1972-79, re-elected 1995
RE536 Hull: Cleaning and Scraping 1968
Etching & aquatint 447 x 496 mm

George Bryan Ingham (1936-1997)
Associate 1972, Fellow 1978

Charles Lloyd (b.1930)
Associate 1972
RE538 Winding into Green 1969
Etching & aquatint 330 x 485 mm

John Higham Grigsby (b.1940)
Associate 1973, Fellow 1978
RE539 Small Hay Spinner 1971
Etching 215 x 433 mm

Leslie Charlotte Benenson (b.1941)
Associate 1974, Fellow 1978
RE560 Charollais Bull II 1973
Wood engraving 146 x 142 mm

André Dunoyer de Segonzac, R.A., Hon. R.W.S. (1884-1974)
Hon. R.E. 1973

Dorothy Trotman Bordass (1905-1992)
Associate 1974, Fellow 1978
RE561 The Monk 1973
Linocut 303 x 230 mm

Olwen Jones, R.W.S. (b.1945)
Associate 1974, Fellow 1978
RE562 Hemlock
Etching & aquatint 525 x 425 mm

Neil McDonald (b.1947)
Associate 1974
RE563 Untitled abstract 1973
Etching & aquatint 224 x 224 mm

Richard Henry Vicary (b.1918)
Associate 1974, Fellow 1992
RE564 "Adaptity" London 1969
Etching & aquatint 250 x 394 mm

James David Beale (b.1948)
Fellow 1975
RE565 Autumn Landscape 1971
Etching & aquatint 494 x 506 mm

Roy Walker (b.1936)
Associate 1975
RE566 Overgrown Lane, Carnello 1972
Etching 415 x 530 mm

Peter William Wickham (b.1934)
Associate 1975, Fellow 1980
RE567 Pazzi Chapel 1970
Relief print 450 x 641 mm

Bernard Carter (b.1920)
Hon. R.E. 1975

Ian Lowe (b.1935)
Hon. R.E. 1975

Sarah Van Niekerk (b.1934)
Associate 1976, Fellow 1982
RE568 A Fair Field Full of Folk
Wood engraving 103 x 148 mm

Glynn David Laurie Thomas (b.1946)
Associate 1976, Fellow 1986
RE569 Welsh Dream 1974
Etching & aquatint 187 x 298 mm

Ian Gavin Howard Grainger (b.1942)
Associate 1977
RE570 Barry Island: a Welsh Seascape 1974
Aquatint 453 x 608 mm

Brian Johnson (b.1939)
Associate 1977
RE571 Origins of Landscape 1976
Etching & aquatint 195 x 287 mm

Paul Harvey Scull (b.1953)
Associate 1977, Fellow 1987
RE572 Closed but not forgotten before the fall 1987
Screenprint 480 x 358 mm

Claire Dalby, R.W.S. (b.1944)
Associate 1978, Fellow 1982
RE573 Old Boat, Flatanger, Norway 1975
Wood engraving 74 x 124 mm

Prof. David Laurence Carpanini, P.R.E., Hon.R.W.S. (b.1946)
Associate 1979, Fellow 1982, President 1995-
RE574 South Winder 1978
Etching 350 x 490 mm

Prof. Michael Kauffmann, Hon. R.W.S. (b.1931)
Hon. R.E. 1977

Prof. Francis Adams Comstock (1897-1981)
Hon. R.E. 1977

Sir Hugh Casson, C.H., K.C.V.O., P.P.R.A., Hon.R.W.S. (b.1910)
Hon. R.E. 1977

Joseph Lawrence Fereday (b.1917)
Associate 1979, Fellow 1984
RE575 Fortress, Les Baux de Provence 1978
Etching & aquatint 376 x 630 mm

Barry Owen Jones, R.W.S. (b.1934)
Associate 1979, Fellow 1982
RE576 Le Trépied, Dolmen 1978
Etching & aquatint 270 x 300 mm

Rosamund Clare Jones (b.1944)
Associate 1979, Fellow 1984
RE577 On my Way to a Party
Etching 282 x 443 mm

Anne Jope (b.1945)
Associate 1979, Fellow 1984
RE578 Three Courtiers 1979
Wood engraving 152 x 83 mm

John A. McPake (b.1943)
Associate 1979, Fellow 1984
RE579 Simon's Herb Garden 1977
Etching & aquatint 411 x 306 mm

William Mann (1914-1979)
Associate 1979
RE580 Tall Ships, Millbay Docks, Plymouth 1977
Wood engraving 128 x 102 mm

Peter John Roach (b.1946)
Associate 1979, Fellow 1982

Vikki Slowe (b.1947)
Associate 1979, Fellow 1982
RE582/1 Pavements 1978
RE582/2 Orbits in Space
Etchings & aquatint 421 x 590; 250 x 337 mm

Brian Holgate Sowerby (1920-1993)
Associate 1979, Fellow 1987
RE583/1 Msida 1978
RE583/2 Japonica 1987
Etchings & aquatint 390 x 540; 76 x 191 mm

Kathe Strenitz (b.1923)
Associate 1979, Fellow 1982
RE584 Railway Shed 1977
Woodcut 418 x 590 mm

Joseph William Winkelman, P.P.R.E., Hon. R.W.S. (b.1941)
Associate 1979, Fellow 1982, President 1989-95
RE585 Hatching Lane, Leafield 1979
Etching 190 x 467 mm

Roger Withington (b.1943)
Associate 1979

Rodney Millard (1921-1980)
Hon. R.E. 1979

Prof. Alan Woodruff, C.M.G. (1916-1992)
Hon. R.E. 1979

Edward Bawden, C.B.E., R.A., Hon. R.W.S. (1903-1989)
Hon. R.E. 1979
RE587 Cabin in the Forest 1954
Engraving 194 x 117 mm

Dorothea Wight (b.1944)
Associate 1979

Mark Balakjian (b.1940)
Associate 1979

Richard Bawden, R.W.S. (b.1936)
Associate 1979, Fellow 1987
RE591 Novodevichy Convent, Moscow
Etching & aquatint 466 x 361 mm

Richard Beer (b.1928)
Associate 1979

Michael Carlo (b.1945)
Associate 1979, Fellow 1991
RE593 Footpath - Winter Morning 1985
Lithograph 316 x 460 mm

Peter Daglish (b.1930)
Associate 1979, Fellow 1992

RE594 Jeudi Matin
Linocut 512 x 402 mm

Phil Greenwood (b.1943)
Associate 1979, Fellow 1982
RE595 Claremont Park 1982
Etching & aquatint 380 x 435 mm

George Worsley Adamson (b.1913)
Associate 1981
RE596 Filming "The Onedin Line" - Exeter 1980
Etching & drypoint 307 x 435 mm

Michael Middleton (b.1950)
Associate 1981, Fellow 1986
RE597 Snow in Barking Park 1987
Etching 136 x 143 mm

Edward Stamp (b.1939)
Associate 1981

Ian Stephens (b.1940)
Associate 1981, Fellow 1984
RE599 Midsummer 1984
Wood engraving 100 x 75 mm

June Berry, R.W.S. (b.1924)
Associate 1982, Fellow 1986
RE600 The Onlookers 1981
Etching & aquatint 311 x 372 mm

James Alexander Boyd (b.1943)
Associate 1982
RE601 Dark Abstract 1981
Etching & aquatint 605 x 490 mm

Eileen Constance Greenwood (b.1915)
Associate 1982, Fellow 1987
RE602/1 Brighton Pavilion
RE602/2 Six Cats of St. Bonnet in Perigord
Etchings 150 x 100; 224 x 296 mm

Karólína Lárusdóttir, R.W.S. (b.1944)
Associate 1982, Fellow 1986
RE603 The Piano Movers Thank the Angel
Etching & aquatint 195 x 292 mm

Frans Wesselman (b.1953)
Associate 1982, Fellow 1986
RE604/1 Atlas 1982
RE604/2 Windmill
Etchings & aquatint 310 x 242; 225 x 223 mm

Carol Elizabeth Walklin (b.1930)
Associate 1982, Fellow 1984
RE605 Girls on Platform 3 1982
Etching 322 x 213 mm

Peter Jeffrey Matthews (b.1942)
Associate 1983, Fellow 1984

RE606 Northern Stack 1982
Etching & aquatint 345 x 494 mm

Colin Frank See-Paynton (b.1946)
Associate 1983, Fellow 1986
RE607/1 Barn Owls 1982
RE607/2 Mute Swan and Carp 1986
Wood engravings 90 x 154; 196 x 248 mm

Norman Ackroyd, R.A. (b.1938)
Associate 1982, Fellow 1984
RE608/1 Tudhill Bushes 1987
RE608/2 Study of Bacchus 1986
Aquatints 423 x 598; 166 x 195 mm

Jo Barry (b.1944)
Associate 1984, Fellow 1986
RE609 Listening to Silence
Etching & aquatint 250 x 290 mm

Carmen Gracia (b.1935)
Associate 1984, Fellow 1987
RE670 Memorial
Intaglio 500 x 440 mm

Stanley Lawrence (1900-1987)
Hon. R.E. 1984

Douglas Wakefield
Associate 1984
RE671 Double Pike 1983
Etching, aquatint & stencil 471 x 243 mm

Tate Adams (b.1922)
Associate 1985, Fellow 1992
RE672 Palm Landscape II
Wood engraving 269 x 281 mm

Alistair Fletcher (b.1963)
Associate 1985, Fellow 1991
RE673/1 Cromlech
RE673/2 Black Shell and Horn
Etchings & aquatint 492 x 586; 243 x 337 mm

Graham Evernden (b.1947)
Associate 1985
RE674 Water Meadows
Etching & aquatint 297 x 236 mm

John Lawrence (b.1933)
Associate 1985, Fellow 1987
RE675/1 Darlington 1978
RE675/2 Nothingmas Day 1985
Wood engravings 232 x 310; 392 x 259 mm

Hilary Paynter (b.1943)
Associate 1985, Fellow 1987
RE676 South Africa 1987
Wood engraving 203 x 247 mm

Richard Shirley Smith (b.1935)
Associate 1985

Julian Trevelyan, R.A. (1910-1988)
Hon. R.E. 1985
RE678 Me and Tom 1976
Etching & aquatint 352 x 475 mm

Michael Rothenstein, R.A. (1908-1993)
Hon. R.E. 1985
RE679 Subway 1985
Woodcut 563 x 758 mm

Leonard Marchant (b.1929)
Associate 1985, Fellow 1986
RE680 Shell & Clock
Mezzotint 278 x 314 mm

Agatha Sorel, A.R.W.S. (b.1935)
Associate 1985, Fellow 1991
RE681 Divine Proportions 1985
Engraving & aquatint 490 x 430 mm

Edwina Ellis (b.1946)
Associate 1985, Fellow 1987
RE682 Mews Topiary 1984
Wood engraving 102 x 64 mm

Sandy Sykes (b.1944)
Associate 1985, Fellow 1987
RE683 Contained
Linocut 650 x 663 mm

Michael Vaughan (b.1938)
Associate 1985
RE684 September 1983
Etching 100 x 137 mm

Simon Brett (b.1943)
Associate 1986, Fellow 1991
RE685 Ramsay Island
Wood engraving 160 x 80 mm

Richard Jebbitt
Associate 1986
RE686 Still Waters 1985
Etching & aquatint 446 x 844 mm

Jane Stobart (b.1949)
Associate 1986, Fellow 1991
RE687 Abbey Mills VI 1985
Etching 234 x 283 mm

Jane Wilson (b.1961)
Associate 1986
RE688/1 The Family
RE688/2 David and Toucan
Etchings & aquatint 529 x 472; 480 x 348 mm

Hilary Adair (b.1943)
Associate 1987, Fellow 1991
RE689 The Lily Pond, Knightshayes 1986
Etching & aquatint 465 x 355 mm

Peter Freeth, R.A. (b.1938)
Associate 1987, Fellow 1991
RE690/1 Kapital City (4)
RE690/2 Shadows over the City 1988
Aquatints 460 x 597; 467 x 600 mm

Peter Ford (b.1937)
Associate 1987, Fellow 1991
RE691 Pumpkins, Gourds and Squashes
Etching & aquatint 375 x 480 mm

Theo Olive (1914-1998)
Associate 1987
RE692 Graduation 1986
Etching & aquatint 251 x 170 mm

Ted Atkinson R.B.S. (b.1929)
Associate 1987, Fellow 1991
RE693 Analogue 1985
Aquatint 340 x 251 mm

Eric Chamberlain (b.1917)
Hon. R.E. 1986

Anthony Dawson (b.1947)
Associate 1987, Fellow 1991
RE694 Fishermen's Huts, Southwold 1986
Etching 79 x 100 mm

Terence Greaves (b.1934)
Associate 1987, Fellow 1991
RE695 Untitled 1985
Aquatint 303 x 373 mm

Paul Hawdon (b.1953)
Associate 1987, Fellow 1991
RE696 Untitled
Etching 487 x 691 mm

Melvyn Petterson (b.1947)
Associate 1987, Fellow 1991
RE697 Hainton Track, Snow 1986
Etching 489 x 388 mm

Jane Anderson (b.1952)
Associate 1988
RE698 A Good Read 1988
Lithograph 405 x 557 mm

Sharon Aivaliotis (b.1951)
Associate 1988, Fellow 1991
RE699 Rosso Veronese 1984
Mezzotint 370 x 543 mm

Brian Hanscomb (b.1944)
Associate 1991, Fellow 1997
RE747 Christ Appears in the Factory 1978-81
Etching 292 x 442 mm

Saša Marinkov (b.1949)
Associate 1992, Fellow 1996
RE748 South-East from Charing Cross 1990
Relief 881 x 581 mm

Trevor Frankland, R.W.S. (b.1931)
Associate 1992, Fellow 1996
RE749 Temple 1992
Linocut 466 x 703 mm

Simon Casson (b.1965)
Associate 1992
RE750 Palm Sunday 1992
Etching & aquatint 249 x 330 mm

Katie Clemson (b.1950)
Associate 1992, Fellow 1996
RE751 Big Red Rock 1991
Relief 590 x 788 mm

Harry Brockway (b.1958)
Associate 1992

Konstantin Kalinovich (b.1959)
Associate 1992, Fellow 1997
RE753 Fish Flying into Double Landscape 1990
Etching & aquatint 131 x 107 mm

Vladan P. Micić
Associate 1992
RE754 Napuśteni Grod 1992
Etching 443 x 533 mm

Toni Martina (b.1956)
Associate 1988, Fellow 1992
RE755/1 Electric Avenue
Etching & aquatint 493 x 580 mm
RE755/2 New York Subway
Etching 450 x 605 mm

David Mortimer-Jones (b.1950)
Associate 1993
RE756 Shoreline, Tyddyn Cynal 1993
Etching & drypoint 194 x 247 mm

Rosemary Benson (b.1948)
Associate 1993, Fellow 1997
RE757 Killer Manillas 1992
Etching 138 x 170 mm

Jacqueline Newell (b.1948)
Associate 1993
RE758 Fire Hydrant 1991
Aquatint & etching 603 x 450 mm

Anne Stevens, M.B.E. (b.1928)
Hon. R.E. 1993

Prof. Pat Gilmour (b.1932)
Hon. R.E. 1993

Sir Philip Manning Dowson, C.B.E., P.R.A., R.I.B.A. (b.1924)
Hon. R.E. 1994

Dr. Anthony Dyson (b.1931)
Hon. R.E. 1994

Silvie Turner (b.1946)
Hon. R.E. 1994

Prof. Leonard McComb, R.A., R.W.S. (b.1930)
Fellow 1994,
RE764 Tulips 1995
Etching 630 x 493 mm

Anthony Davies (b.1947)
Fellow 1994
RE765 The Great Divide I 1987
Lithograph 623 x 816 mm

Jim Anderson (b.1965)
Associate 1994, Fellow 1997
RE766 Narcissi 1993
Linocut 743 x 532 mm

Mandy Bonnell (b.1957)
Associate 1994, Fellow 1997
RE767 Lamu No.17 1993
Etching & collage 180 x 180 mm

Roger Harris (b.1942)
Associate 1994
RE768 Peter and the Wolf (The Music of What Happens)
Aquatint & stencil 250 x 214 mm

Jason Hicklin (b.1966)
Associate 1994
RE769 Brecon Beacons from Llanbury 1994
Etching & aquatint 378 x 473 mm

Magnus Irvin (b.1952)
Associate 1994, Fellow 1997
RE770 After Dark
Etching & aquatint 165 x 132 mm

Chippa Sudhakar (b.1967)
Associate 1994
RE771 Dream 1993
Etching & stencil 330 x 246 mm

Prof. James Todd (b.1937)
Associate 1994, Fellow 1997
RE772 George Grosz with Boxer 1984
Woodcut 303 x 405 mm

**Elizabeth Blackadder, O.B.E., R.A.,
Hon.R.W.S., R.S.A.** (b.1931)
Hon. R.E. 1994
RE773 Untitled
Intaglio 430 x 528 mm

Paula Rego (b.1935)
Hon. R.E. 1994

Paul Kershaw (b.1949)
Associate 1994
RE775 Coire Lagan
Wood engraving 125 x 154 mm

Pavel Makov (b.1958)
Associate 1994
RE776 Morning
Etching 344 x 1010 mm

Tiffany McNab (b.1964)
Associate 1995
RE777 Topiary with Curious Gardener
Etching & aquatint 493 x 419 mm

Trevor Price (b.1966)
Associate 1995
RE778 A Twist in the Tail (Proverbs series)   1992
Etching 187 x 245 mm

John Brunsdon (b.1933)
Associate 1995
RE779 Derbyshire Peak Farm
Etching & aquatint 452 x 600 mm

Linda Anne Landers (b.1959)
Associate 1995
RE780 Herald Angels 1994
Wood engraving 165 x 99 mm

Max Werner (b.1956)
Associate 1995
RE781 A Moment of Tension in Battersea 1993
Etching & aquatint 391 x 598 mm

Charles Beauchamp (b.1949)
Associate 1995
RE782 Cause and Effect II 1993
Drypoint 58 x 98 mm

Peter Dover (b.1954)
Associate 1995
RE783 Casares 1994
Relief 380 x 365 mm

Dr. Irvine Stewart Lees Loudon (b.1924)
Associate 1995
RE784 Thorn Tree II 1995
Etching & aquatint 198 x 234 mm

David Case (b.1943)
Hon. R.E. 1995

Craig Hartley (b.1959)
Hon. R.E. 1995

David Landau (b.1950)
Hon. R.E. 1995

Jim Westergard (b.1939)
Associate 1996
RE788 Trespass 1989
Wood engraving 302 x 227 mm

Judith Jaidinger (b.1941)
Associate 1996
RE789 Coup de Grâce
Wood engraving 210 x 287 mm

Naohiko Watanabe (b.1966)
Associate 1996
RE790 Untitled 1996
Silkscreen 685 x 685 mm

Dale Devereux Barker (b.1962)
Associate 1996
RE791 Diary of a Has-been 1995
Reduction lino 246 x 244 mm

Martin Langford (b.1970)
Associate 1996
RE792 Capital Growth 1996
Mezzotint 520 x 423 mm

Ursula Leach (b.1947)
Associate 1996
RE793 Barging Form
Aquatint 602 x 671 mm

Emma Stibbon (b.1962)
Associate 1996
RE794 Botallack Head 1994
Wood engraving 772 x 780 mm

John Doyle, M.B.E., P.R.W.S. (b.1928)
Hon. R.E. 1996

Sandra Baxter (b.1968)
Associate 1997
RE796 Rest
Etching 593 x 468 mm

Jeff Clarke (b.1935)
Associate 1997
RE797 Blown Rose 1993
Etching & aquatint 225 x 170 mm

Vladimir Filipenko (b.1945)
Fellow 1997
RE798 Sad Clown
Woodcut 612 x 770 mm

Vijay Kumar (b.1947)
Fellow 1997
RE799 Print from India Portfolio
Etching & photo-etching 192 x 250 mm

Adrian Bartlett (b.1939)
Associate 1997
RE800 Sisters
Etching 302 x 405 mm

Robert Kipniss (b.1931)
Associate 1998
RE801 Vase, Chair & Trees 1995
Mezzotint 176 x 137 mm

Janet Brooke (b.1947)
Associate 1998
RE802 Street Corner
Etching 418 x 621 mm

Oona Grimes (b.1957)
Associate 1998
RE803 The Big Monster Secret 1996
Etching 595 x 1185 mm

John Purcell (b.1948)
Hon. R.E. 1998

Peter Coker, R.A. (b.1926)
Hon. R.E. 1998

Peter Blake, C.B.E., R.A., Hon.R.W.S. (b.1932)
Hon. R.E. 1998

Alun Leach-Jones (b.1937)
Hon. R.E. 1999

Michael Fell (b.1939)
Hon. R.E. 1999

Mychael Barratt (b.1961)
Associate 1999
RE809 The Great Globe Itself
Etching & aquatint 453 x 378 mm

Daphne C. Casdagli (b.1946)
Associate 1999
RE810 Scarecrows at B.A. France 1996/7
Monoprint-lithograph 374 x 476 mm

David Lintine (b.1957)
Associate 1999
RE811 Interior with Table
Etching 336 x 291 mm

Gennady Pugachevsky (b.1966)
Associate 1999
RE812 The Frog 1997
Engraving 63 x 88 mm

Mary Stewart (b.1937)
Associate 1999
RE813 Figs in a Red Lacquer Bowl 1996
Linocut 457 x 457 mm

Judy Willoughby (b.1949)
Associate 1999
RE814 Rainy Day Pears 1998
Aquatint 394 x 497 mm

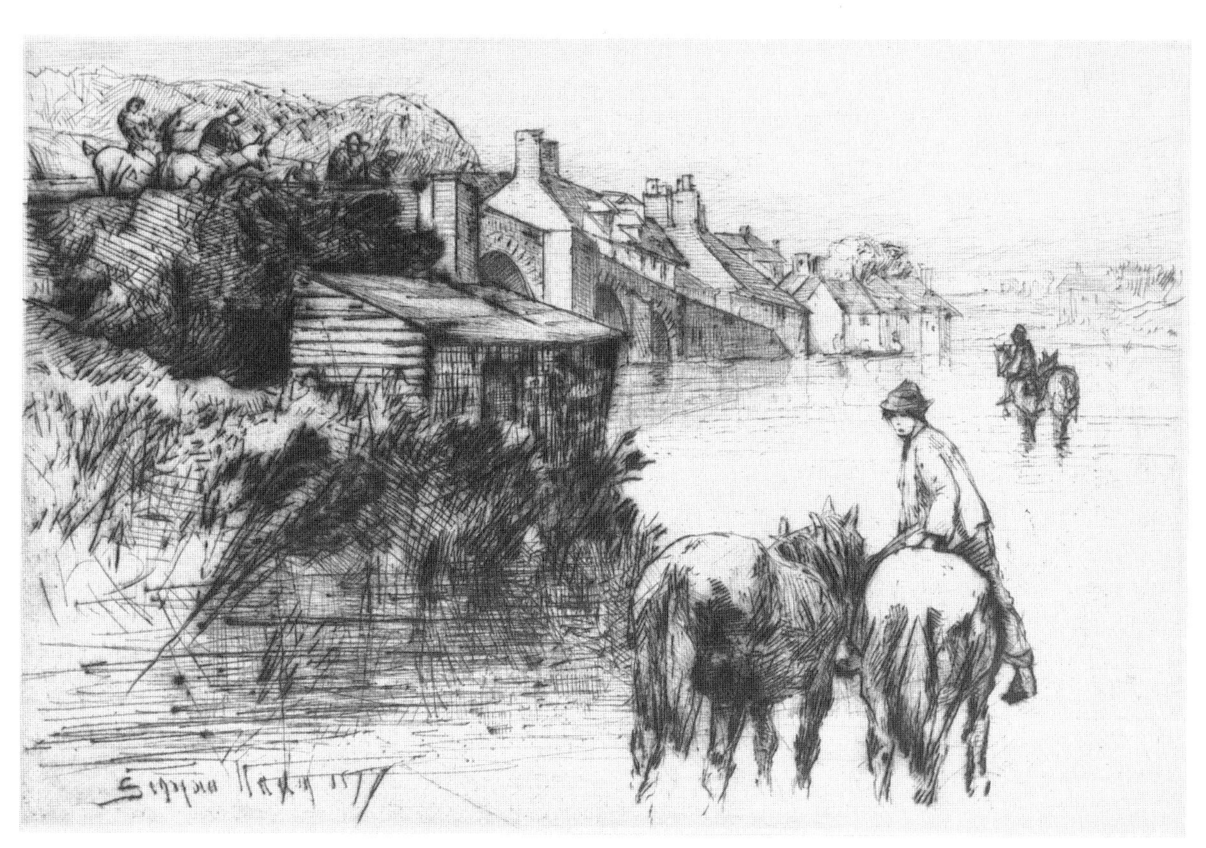

Sir Francis Seymour Haden, M.D., P.R.E. (1818-1910)   RE1 *Wareham Bridge* 1877   Drypoint 149 x 226 mm

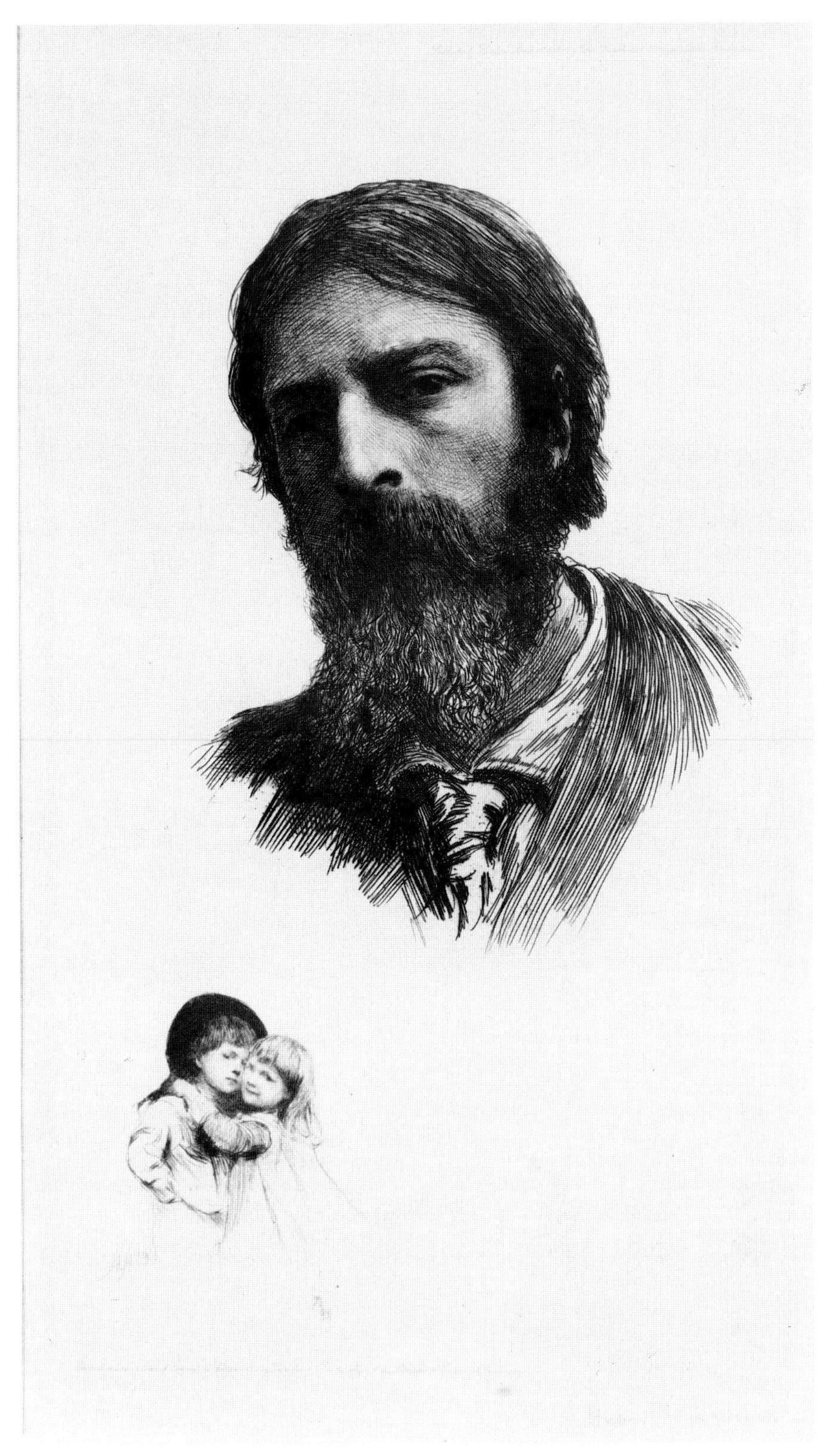

Sir Hubert von Herkomer, C.V.O., R.A., R.W.S. (1849–1914)
RE3 *Self Portrait with his Children* 1879   Etching 352 x 204 mm

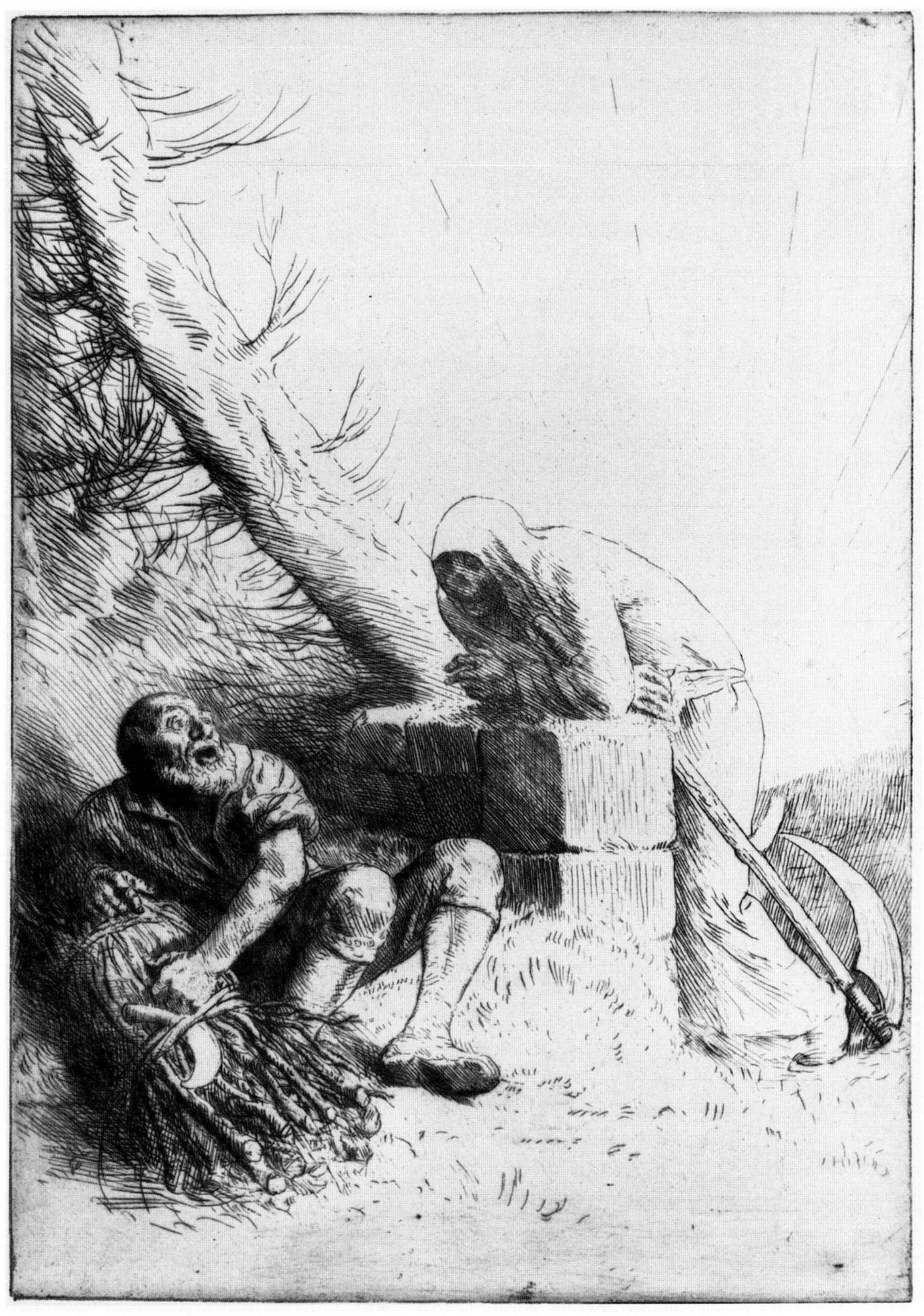

Alphonse Legros (1837–1911)   RE4 *Death and the Woodman* c.1875   Etching 373 x 270 mm

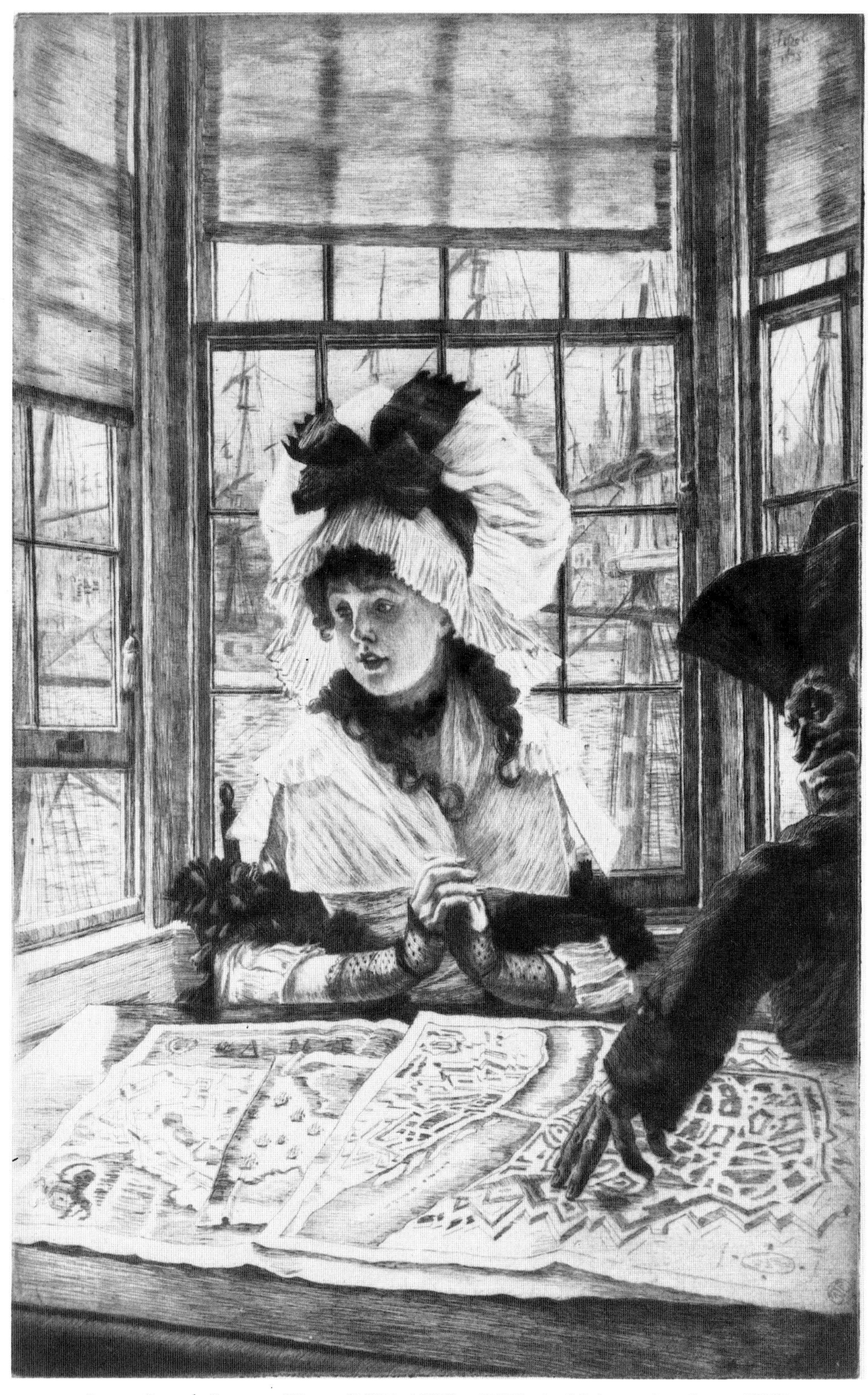

James Joseph Jacques Tissot (1836–1902) RE7 *An Uninteresting Story* 1878
Etching & drypoint 314 x 203 mm

86

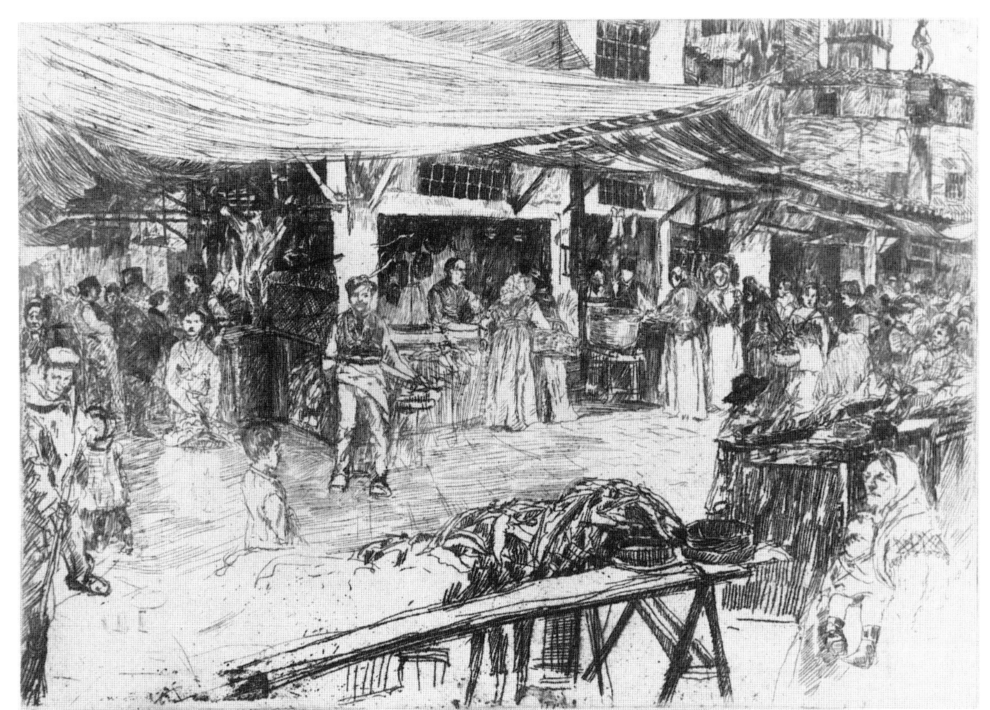

Otto Henry Bacher (1856–1909)   RE20 *Market, Florence* 1880   Etching & drypoint 181 x 257 mm

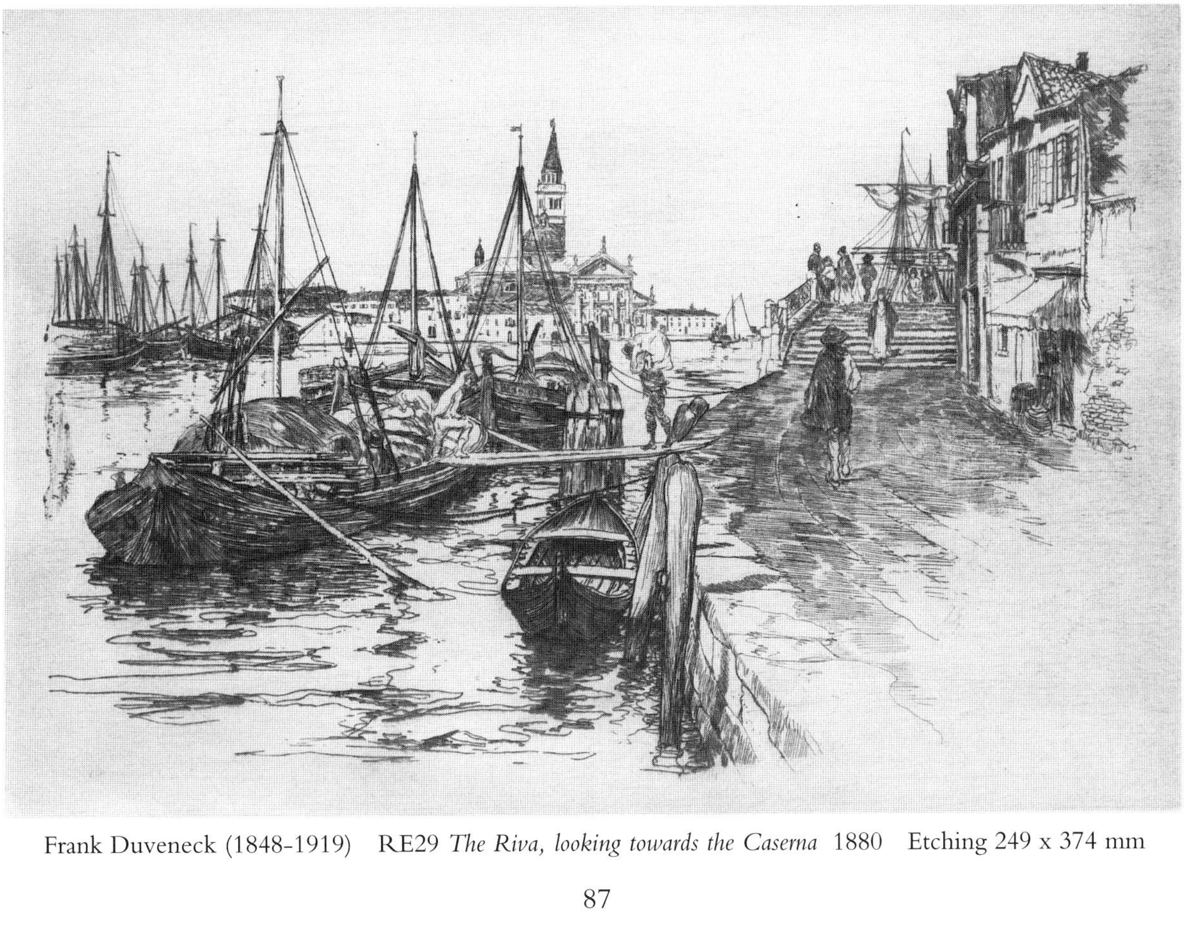

Frank Duveneck (1848–1919)   RE29 *The Riva, looking towards the Caserna* 1880   Etching 249 x 374 mm

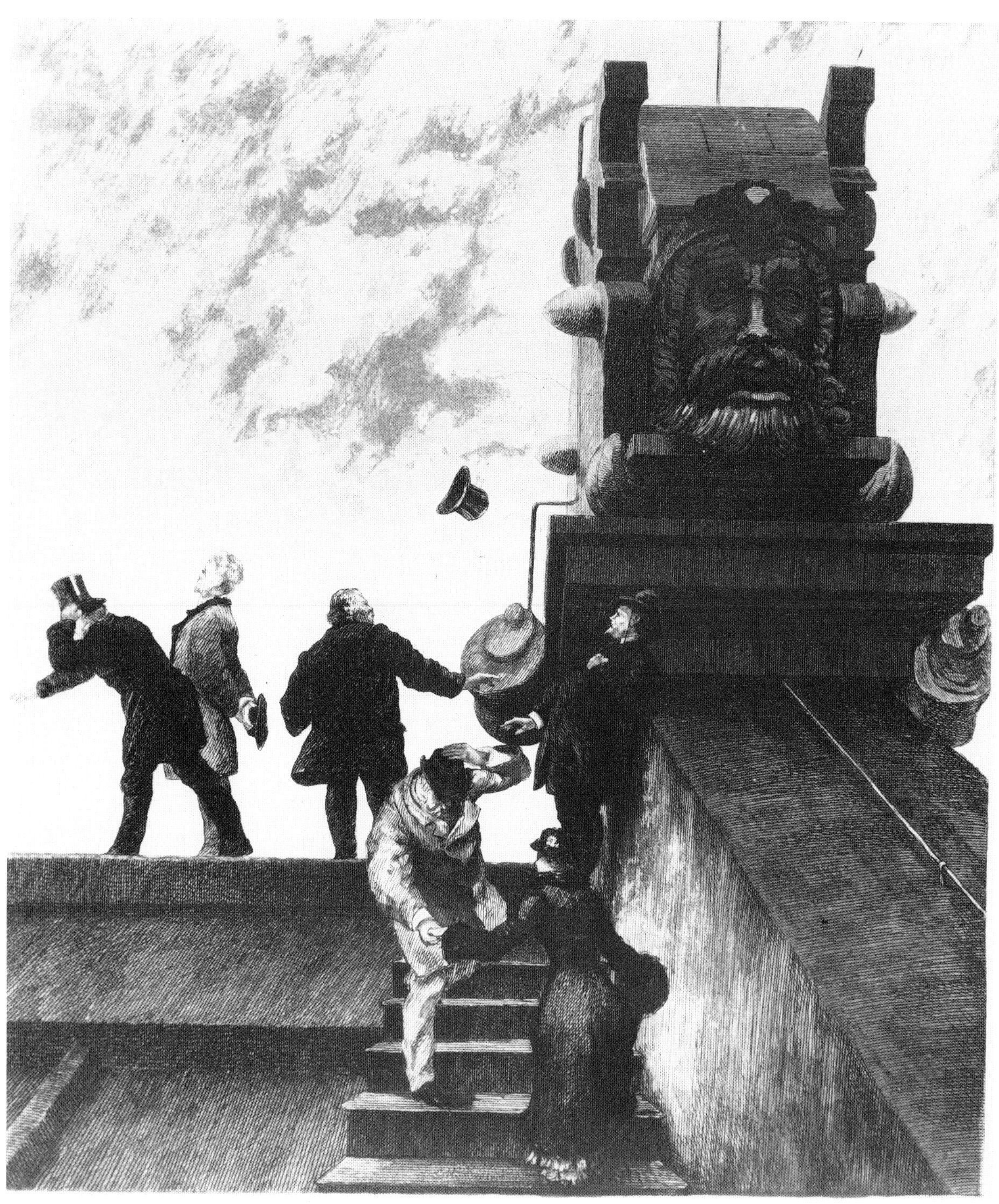

Charles Paul Renouard (1845-1924)   RE62 *Une Visite sur les Toits de l'Opéra à Paris* c.1880–81
Etching & aquatint 222 x 193 mm

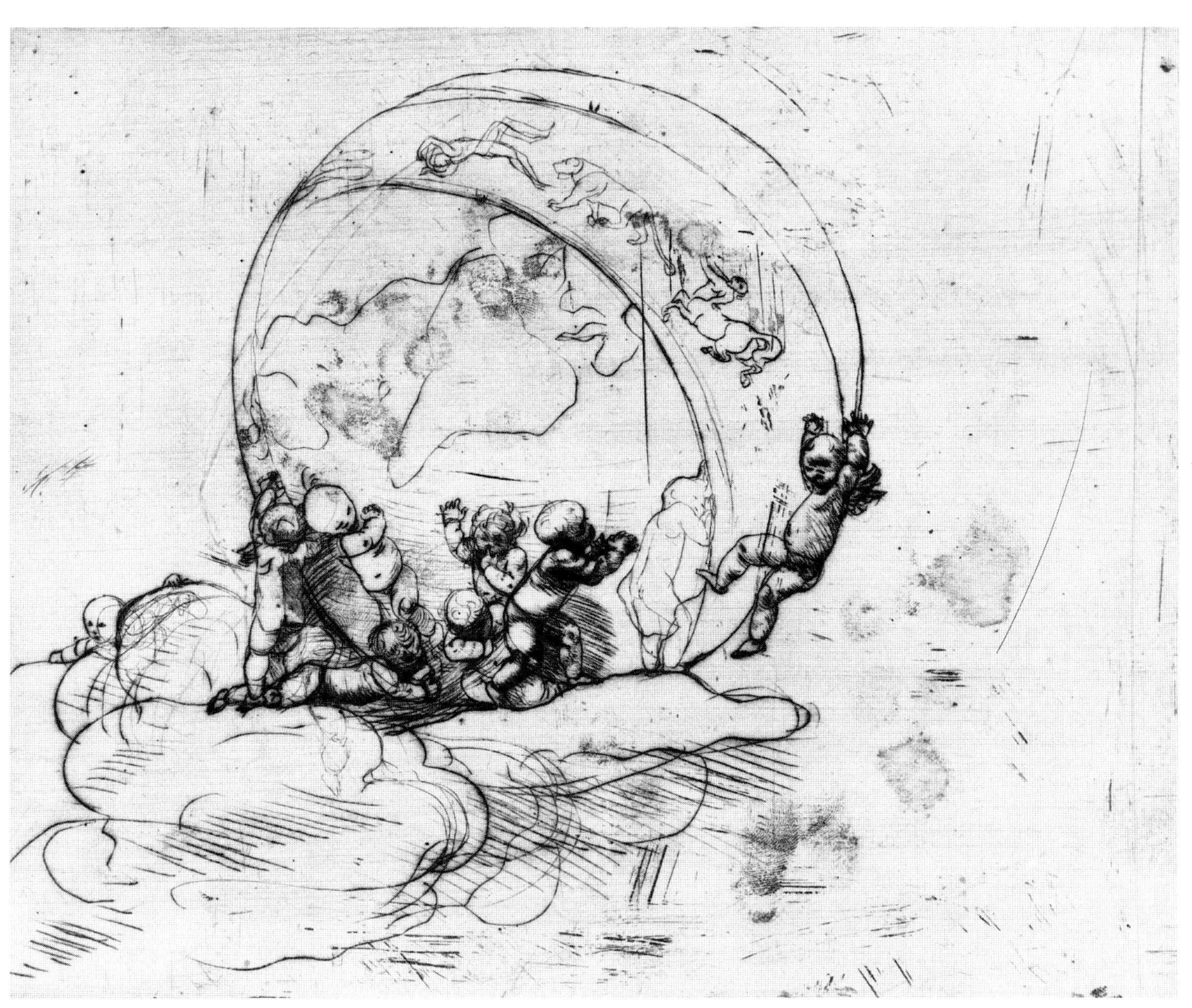

François Auguste René Rodin (1840–1917)   RE87 *A Sketch* 1881   Etching 200 x 248 mm

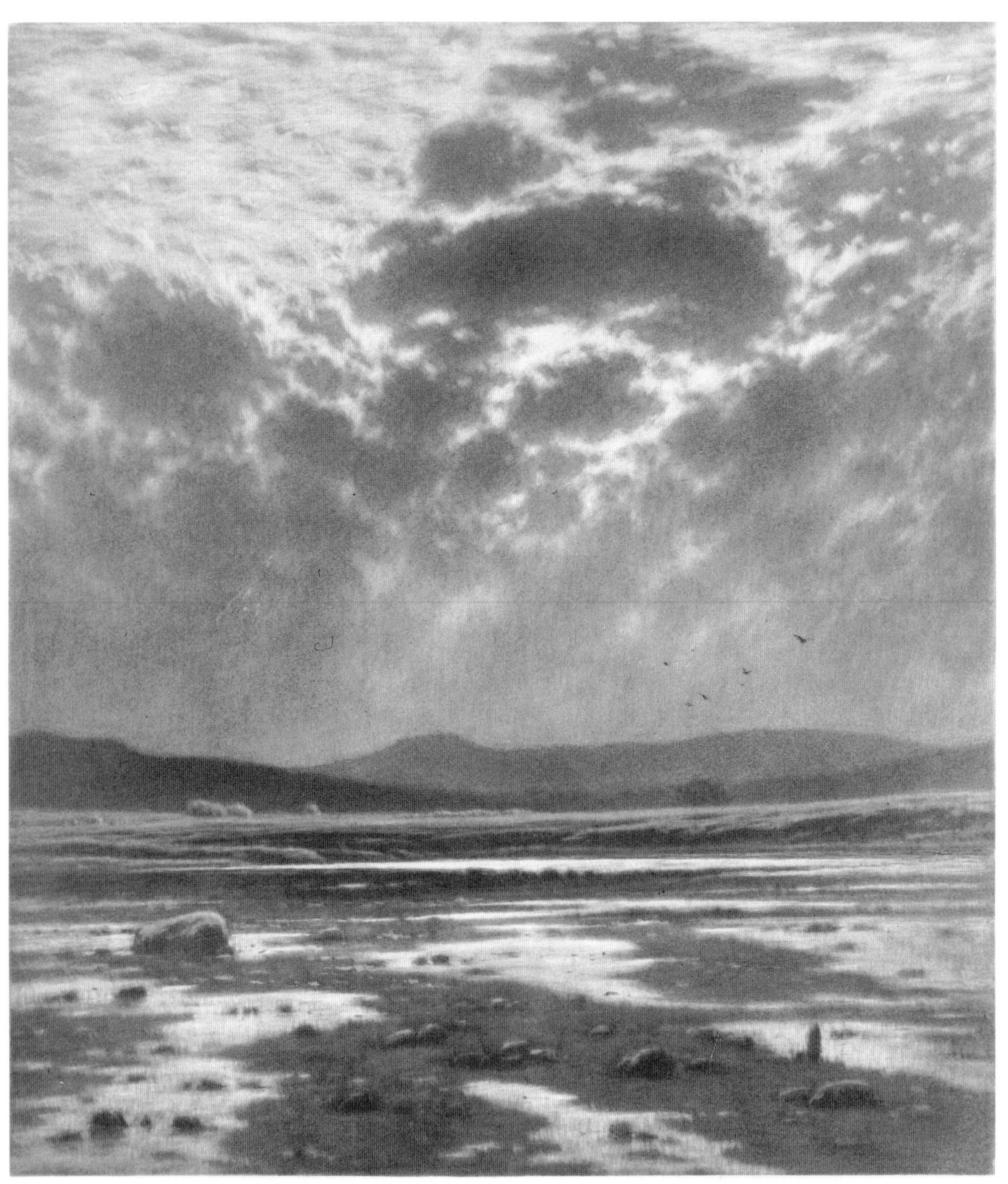

Joseph Knight (1837–1909)   RE92 *Near Barmouth* 1882   Mezzotint 338 x 296 mm

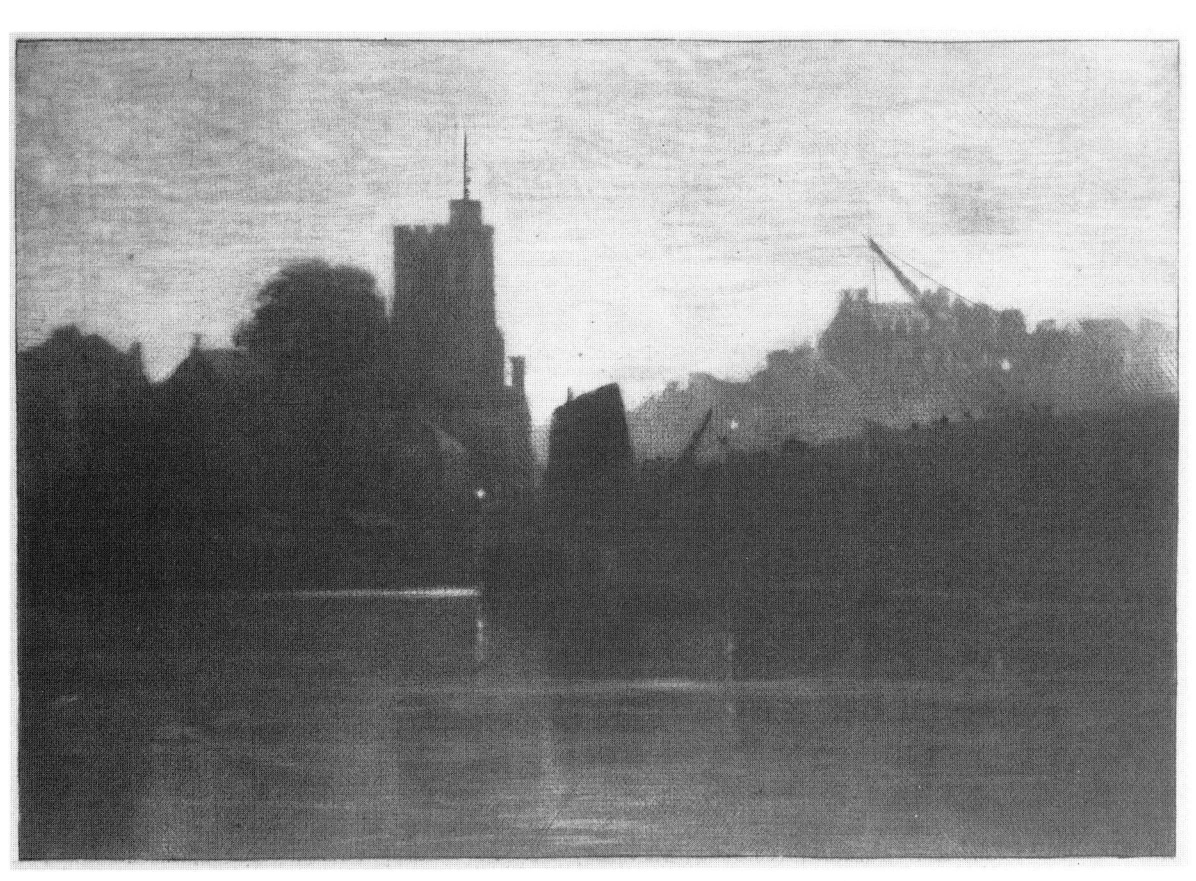

Sir Francis Job Short, P.R.E., R.A. (1857–1945)   RE112 *The Tide ebbs, Putney Bridge* 1885
Mezzotint 125 x 181 mm

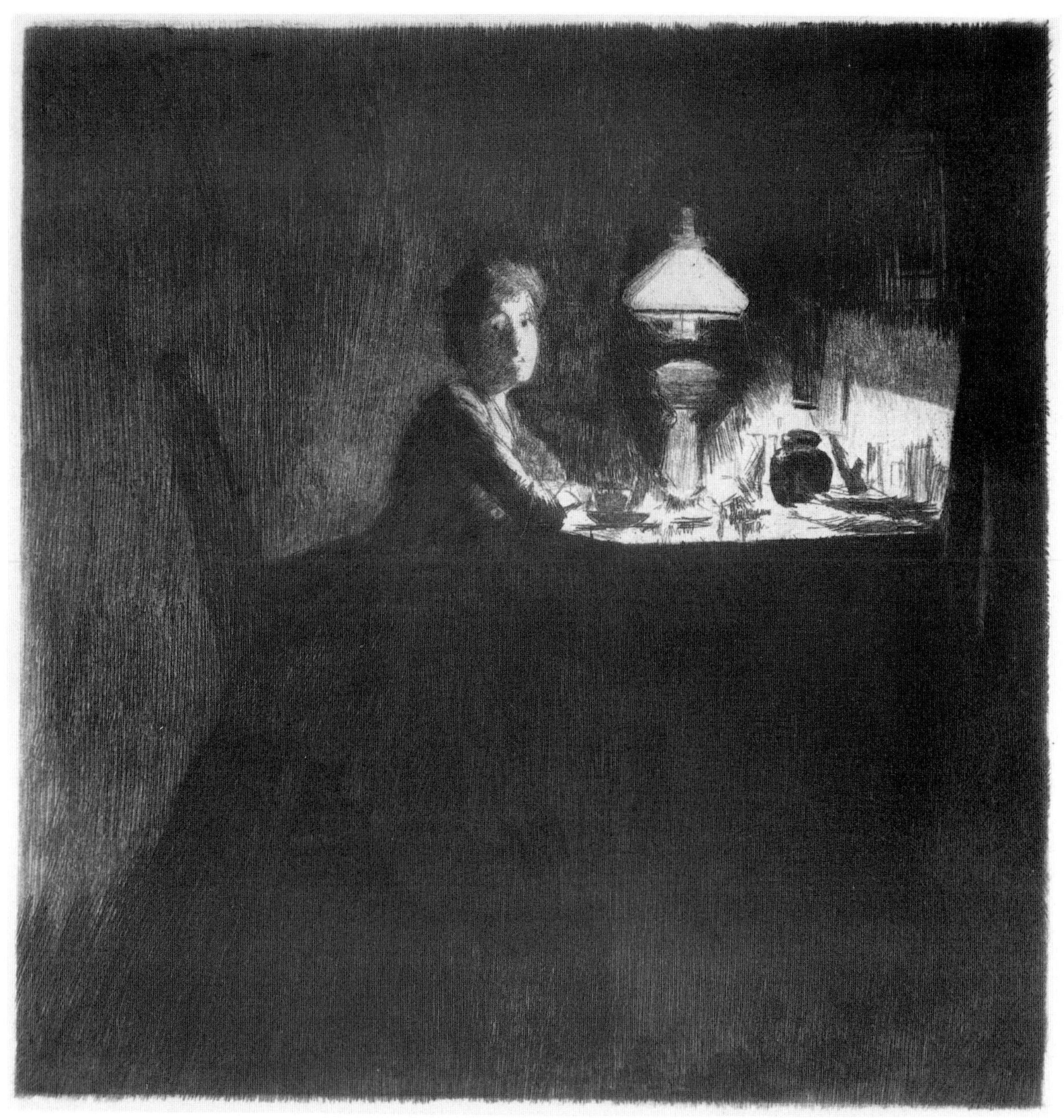

Walter Richard Sickert, R.A. (1860-1942)   RE118 *Louie* 1884   Etching 228 x 225 mm

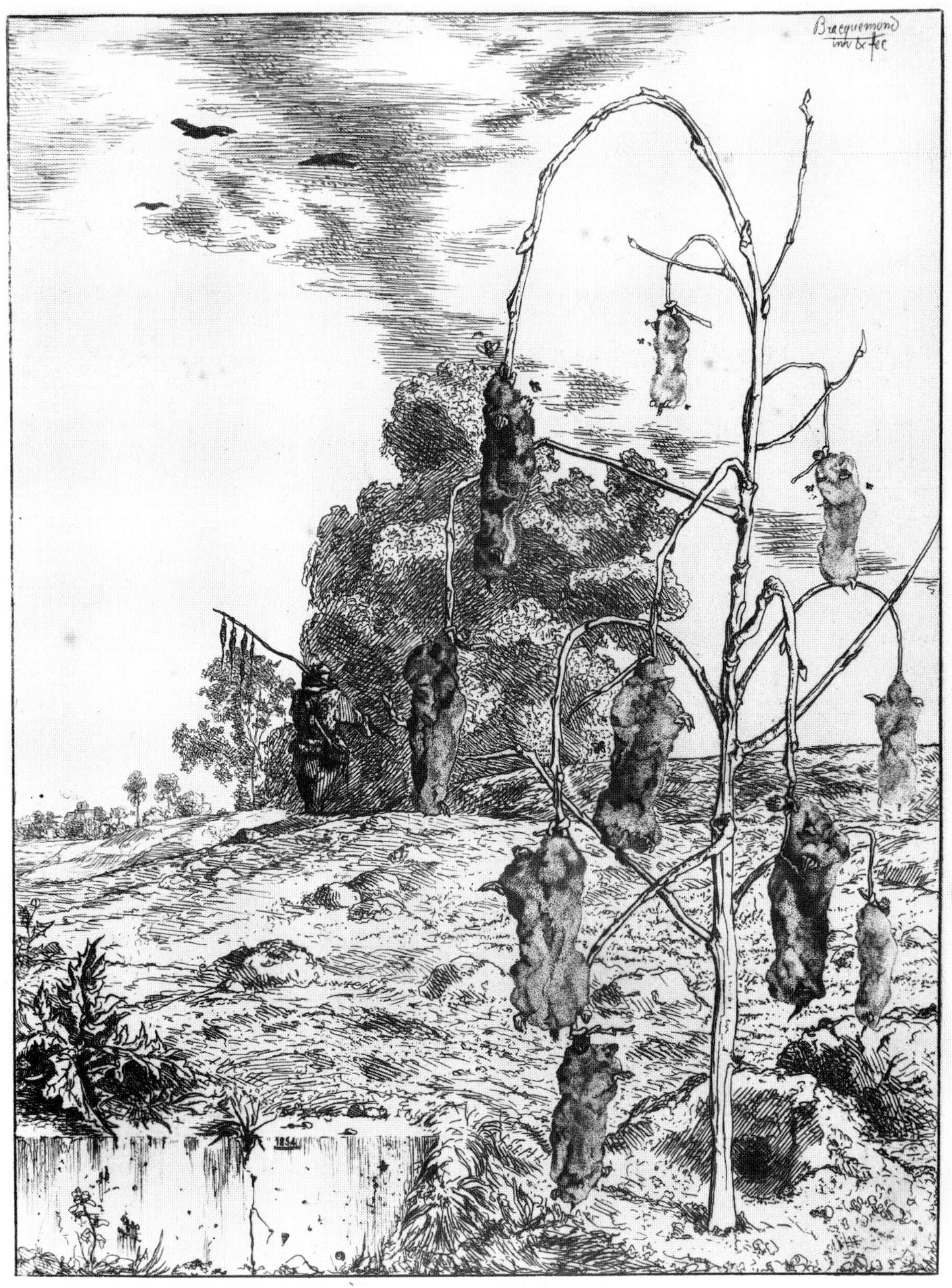

Félix Henri Bracquemond (1833-1914)   RE141 *The Mole Catcher* 1854   Etching 250 x 190 mm

93

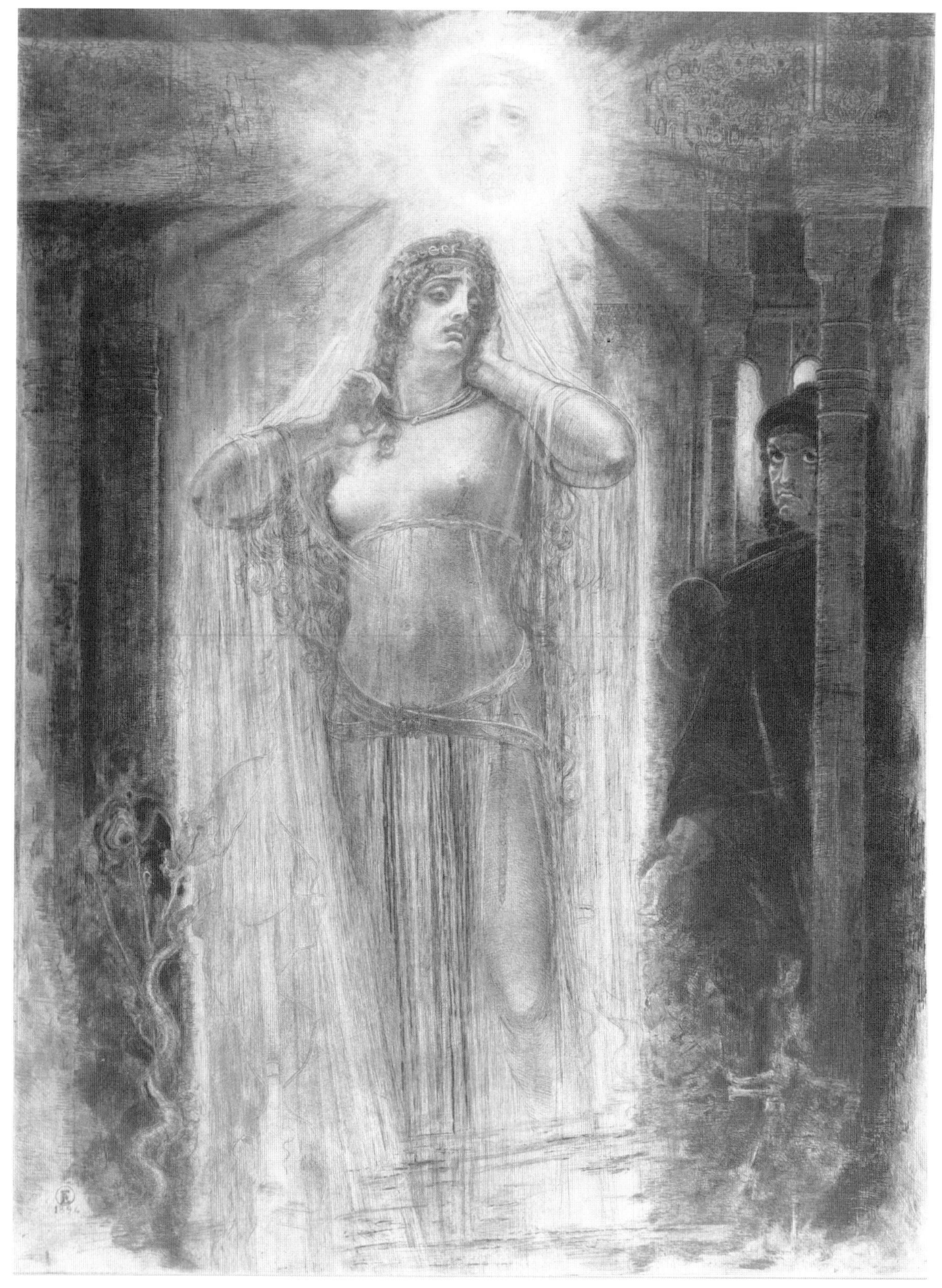

Rogelio de Egusquiza (1845–1915)   RE189 *From Wagner's "Parsifal"* 1894   Etching & drypoint 473 x 350 mm

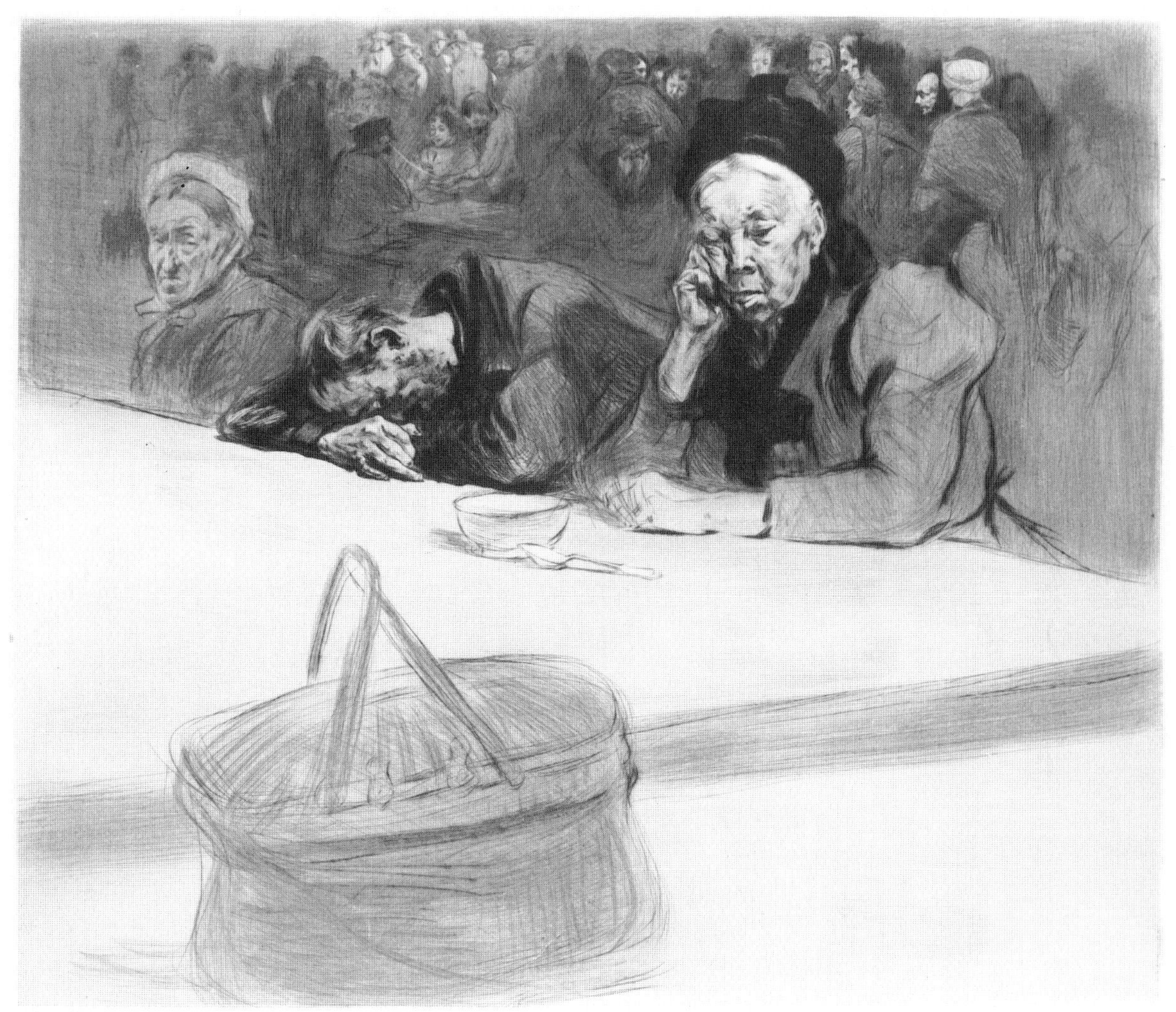

Edgar Chahine (1874–1947)　RE201 *Au Château Rouge*　1899　Drypoint 280 x 334 mm

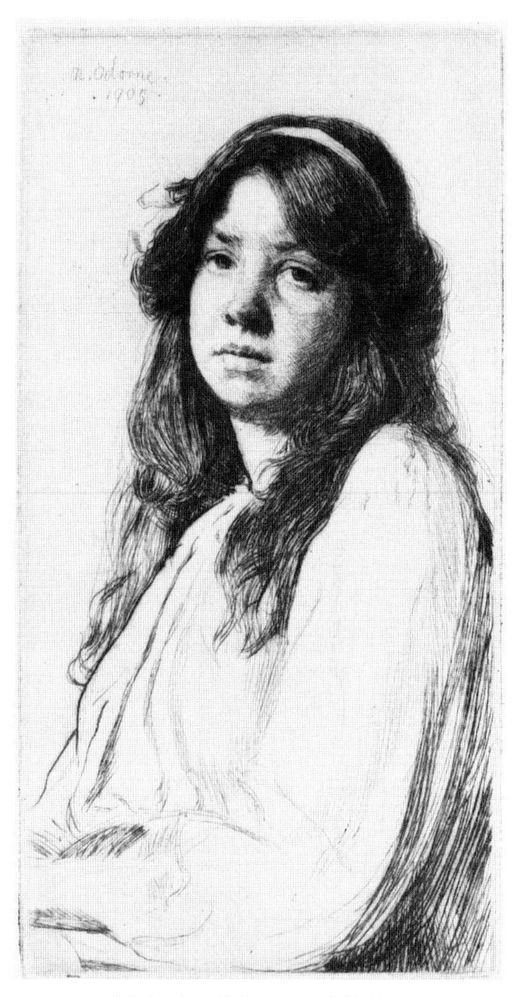

Malcolm Osborne, C.B.E.,
P.R.E., R.A. (1880–1963)
RE230 *Maggie, Study of a Girl's Head* 1905
Etching 127 x 65 mm

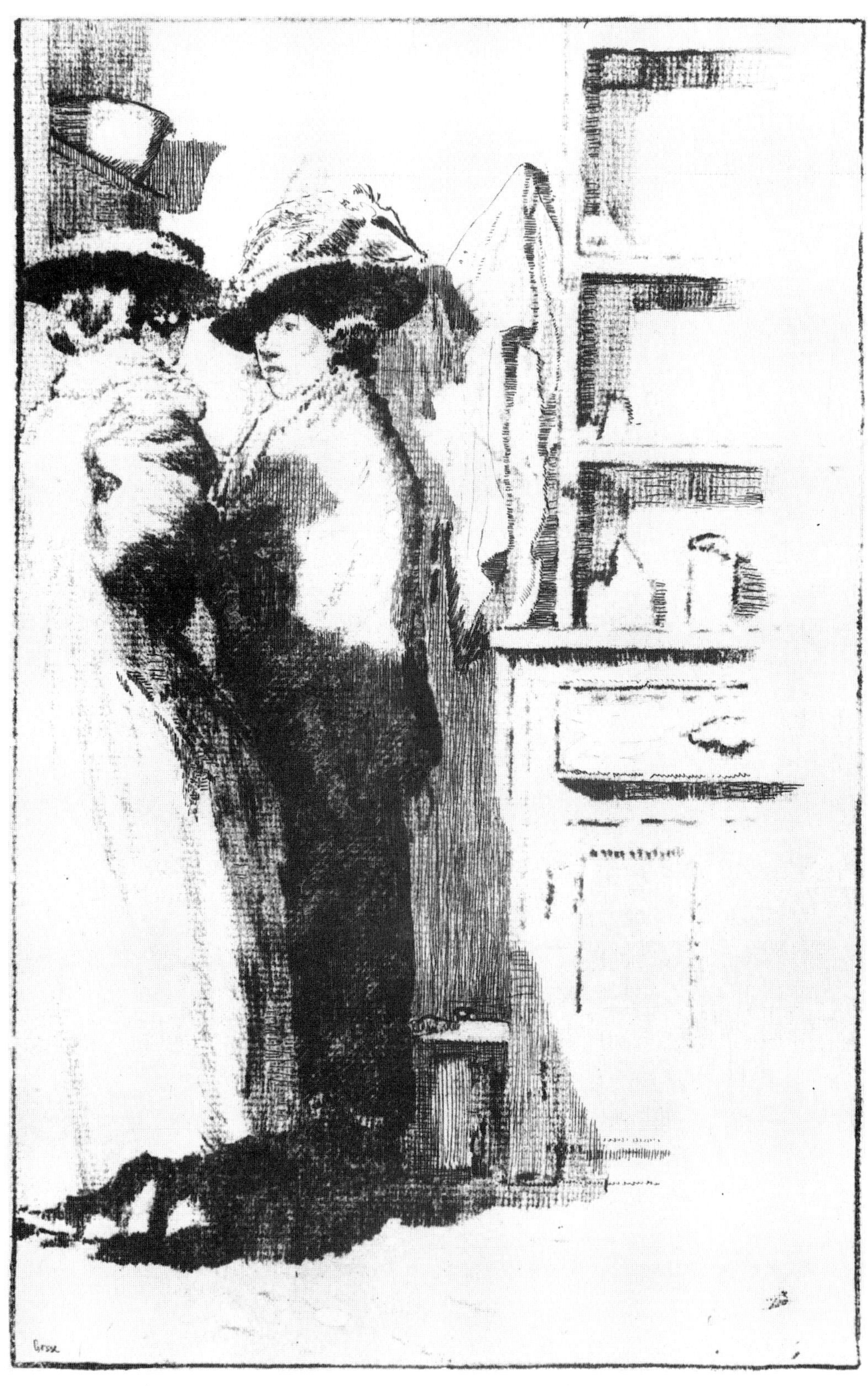

Laura Sylvia Gosse (1881–1968)   RE286 *Costers by Sink*   Etching 234 x 152 mm

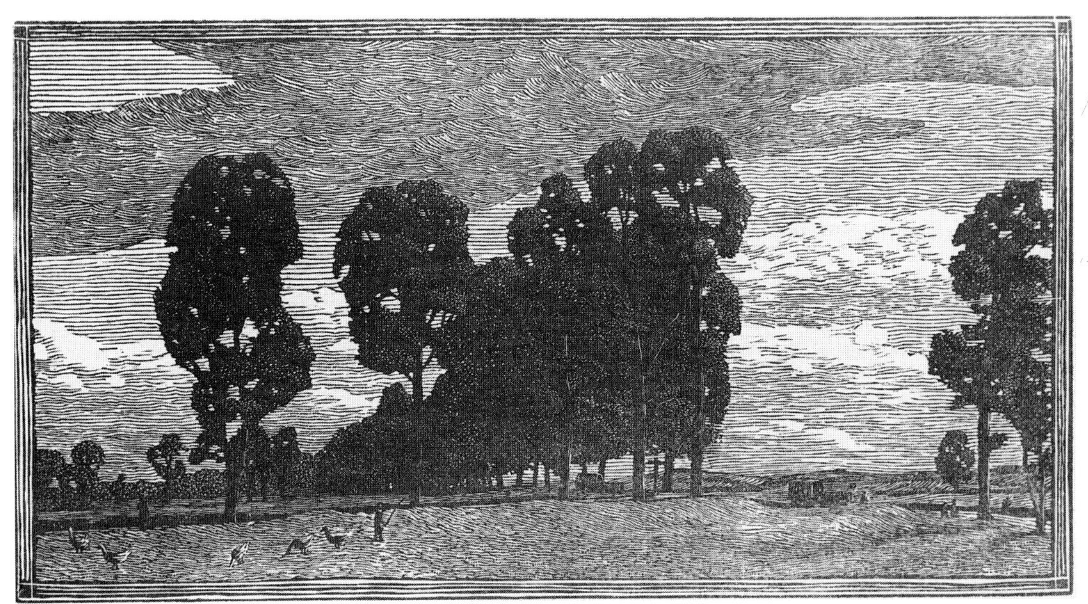

Gwendolen Raverat (1885-1957)   RE302 *The Gooseherd* 1919   Wood engraving 101 x 191 mm

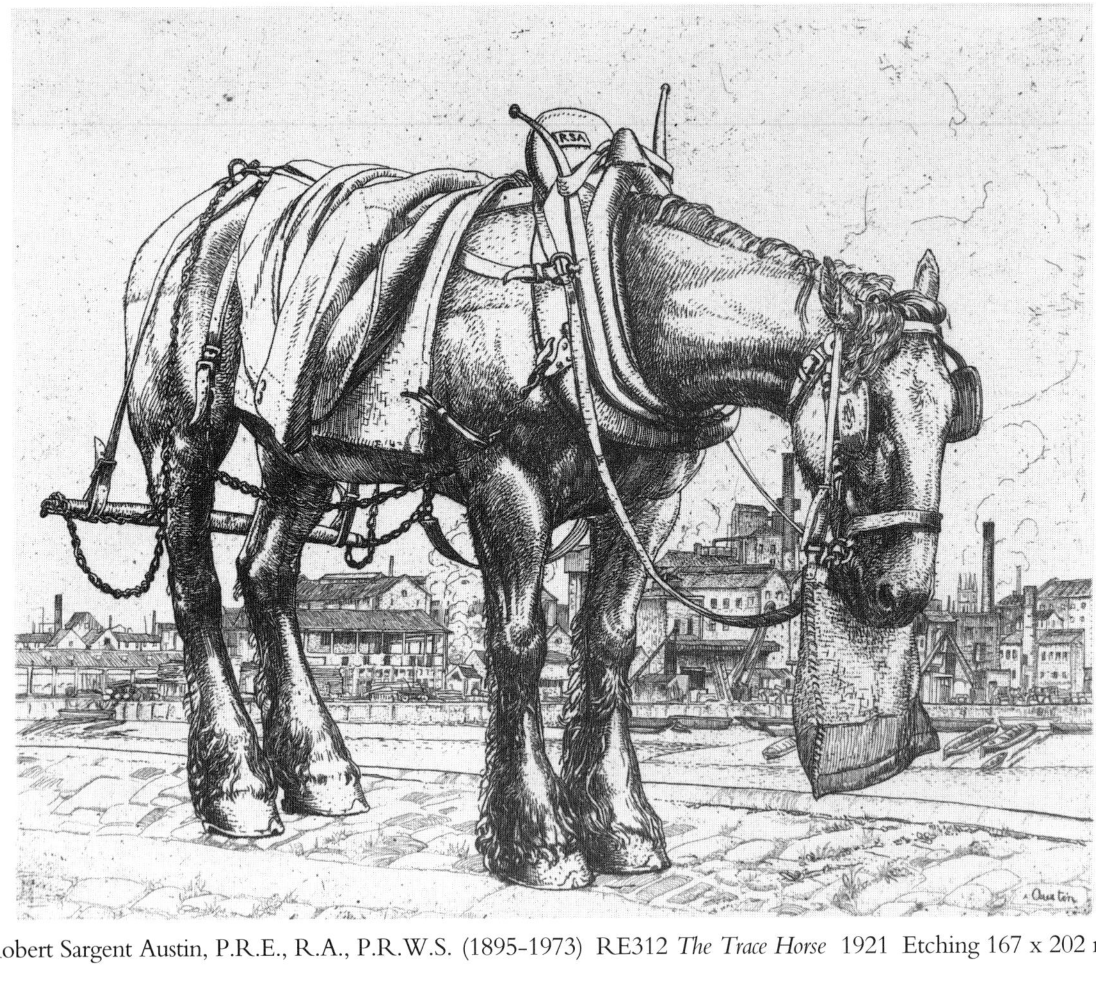

Robert Sargent Austin, P.R.E., R.A., P.R.W.S. (1895-1973) RE312 *The Trace Horse* 1921 Etching 167 x 202 mm

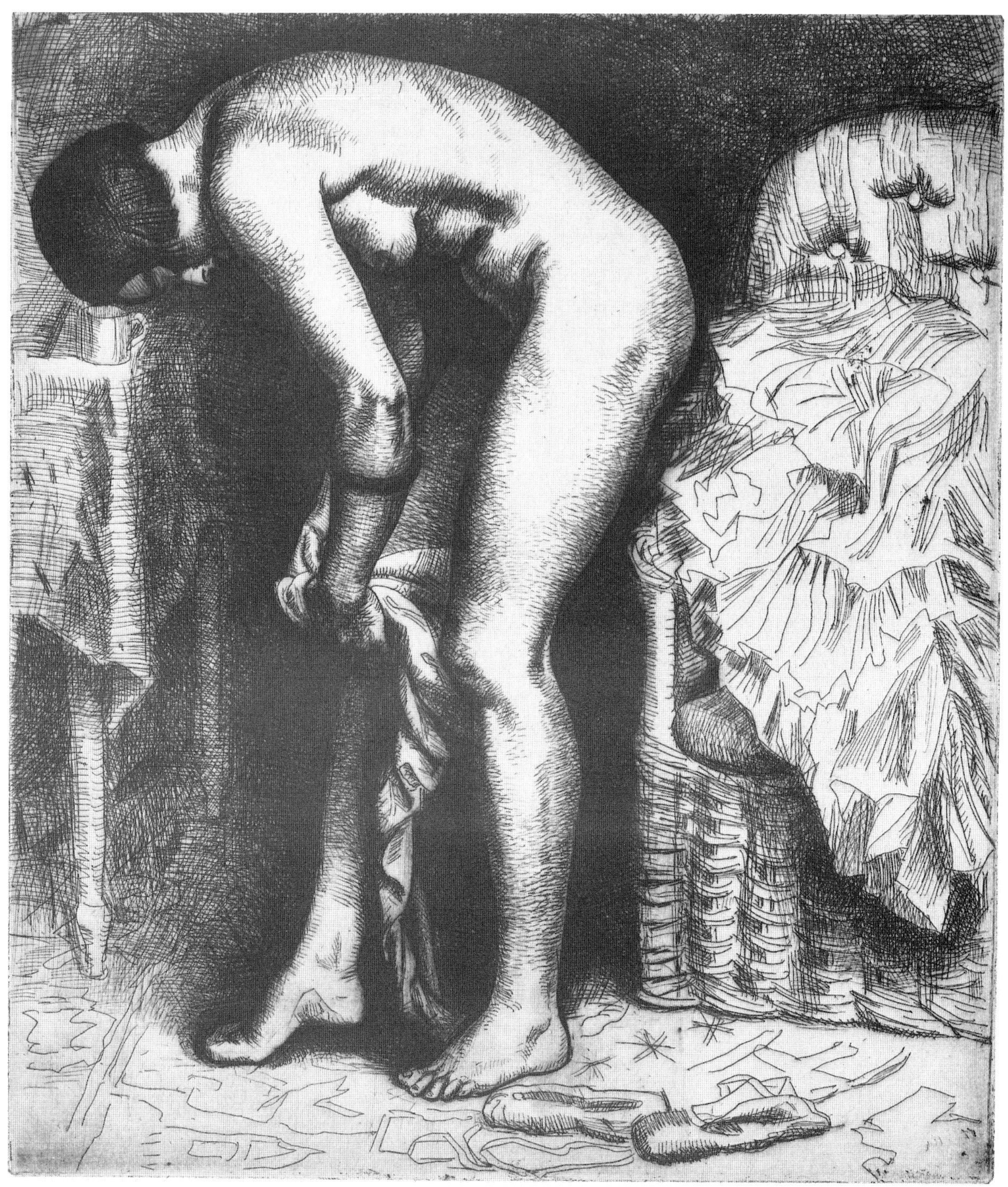

Dame Laura Knight, D.B.E., R.A., R.W.S. (1877–1970)   RE326 *Putting on Tights* 1926   Etching 202 x 176 mm

99

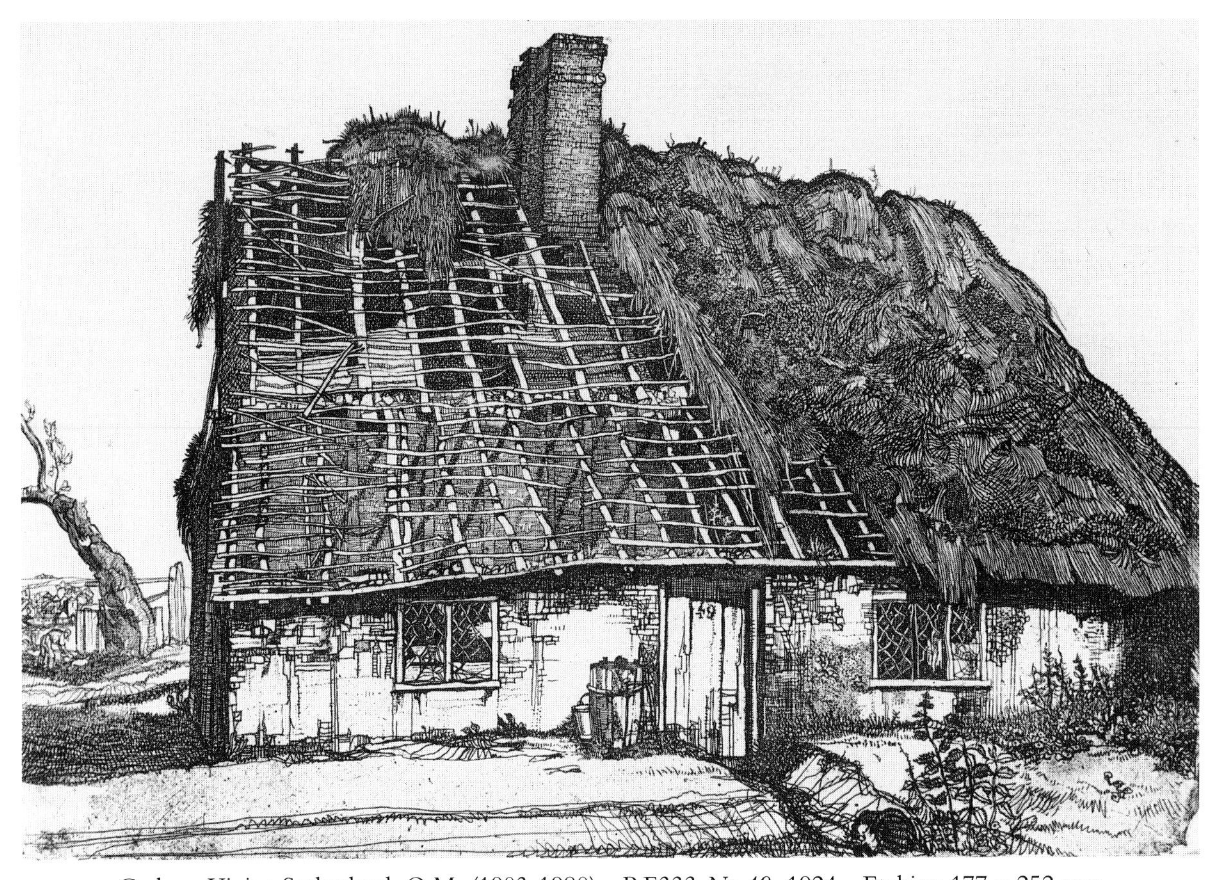

Graham Vivian Sutherland, O.M. (1903–1980)   RE333 *No. 49* 1924   Etching 177 x 252 mm

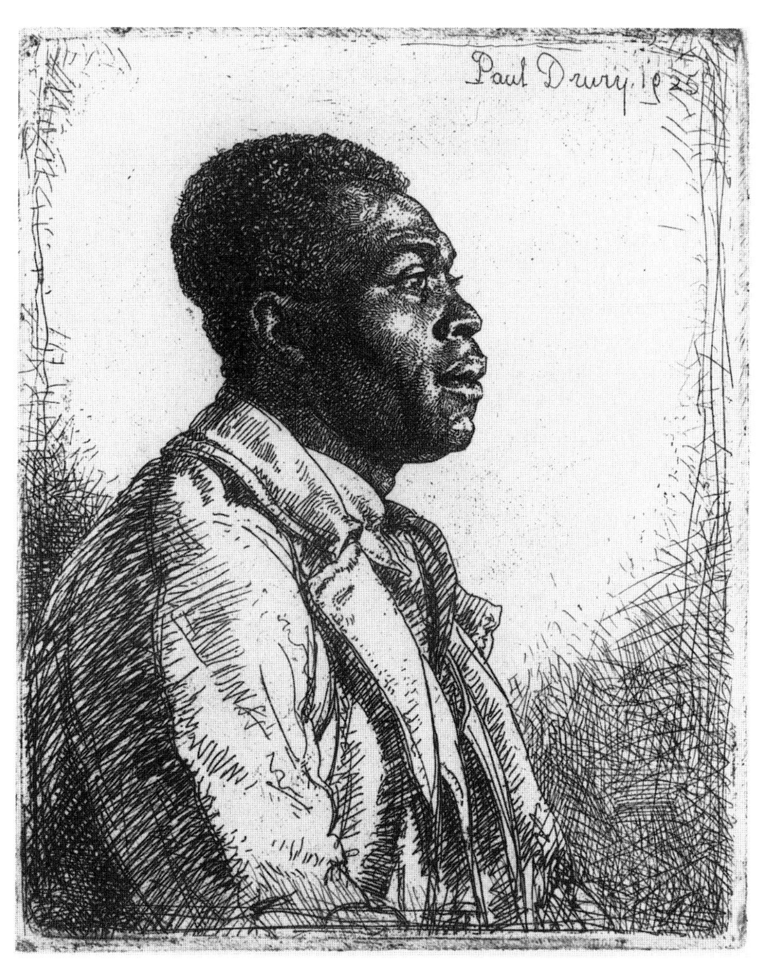

Paul Dalou Drury, P.R.E. (1903–1987)   RE337 *Negro* 1925
Etching 123 x 100 mm

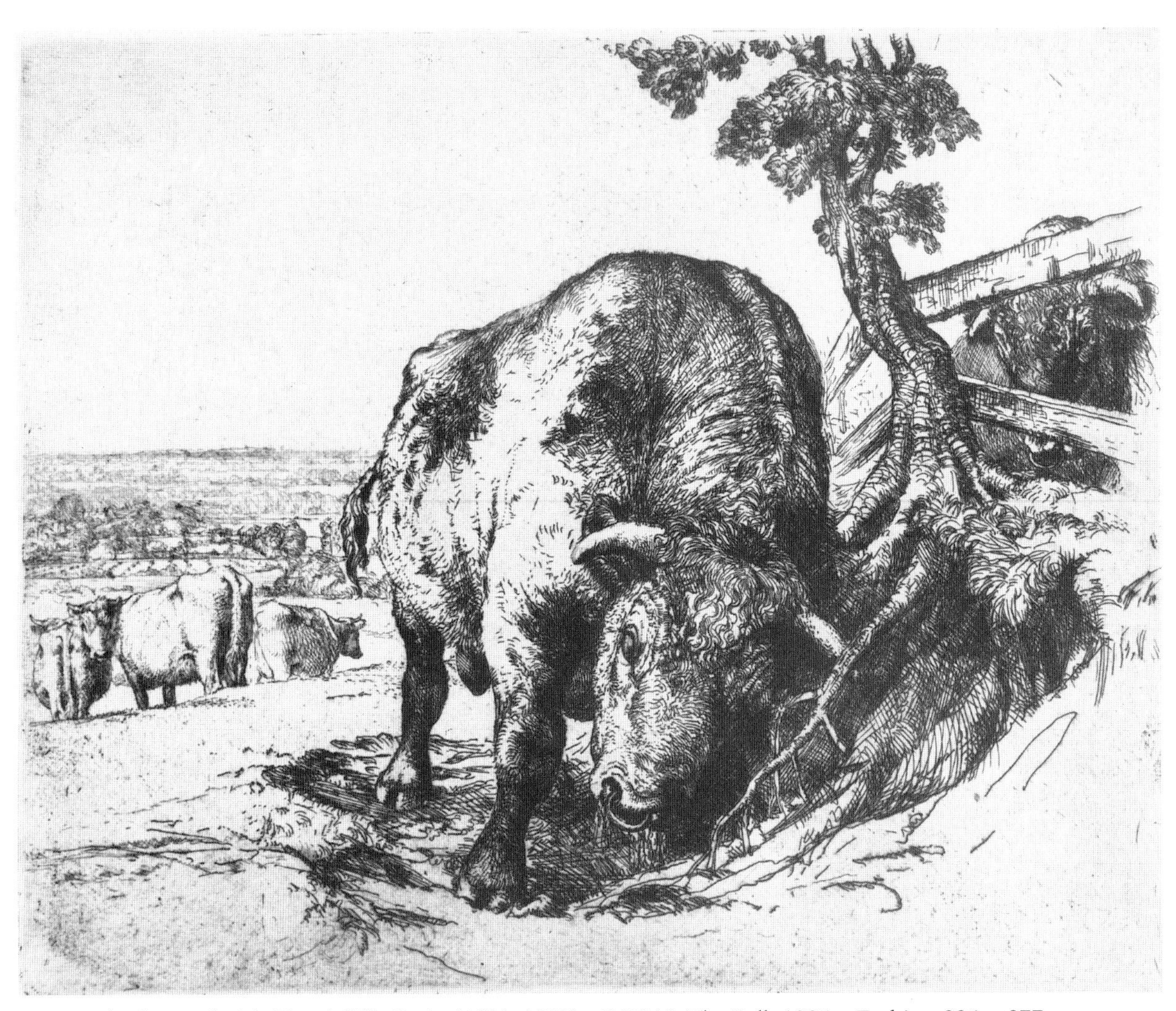

Charles Frederick Tunnicliffe, R.A. (1901–1979)   RE343 *The Bull* 1926   Etching 226 x 277 mm

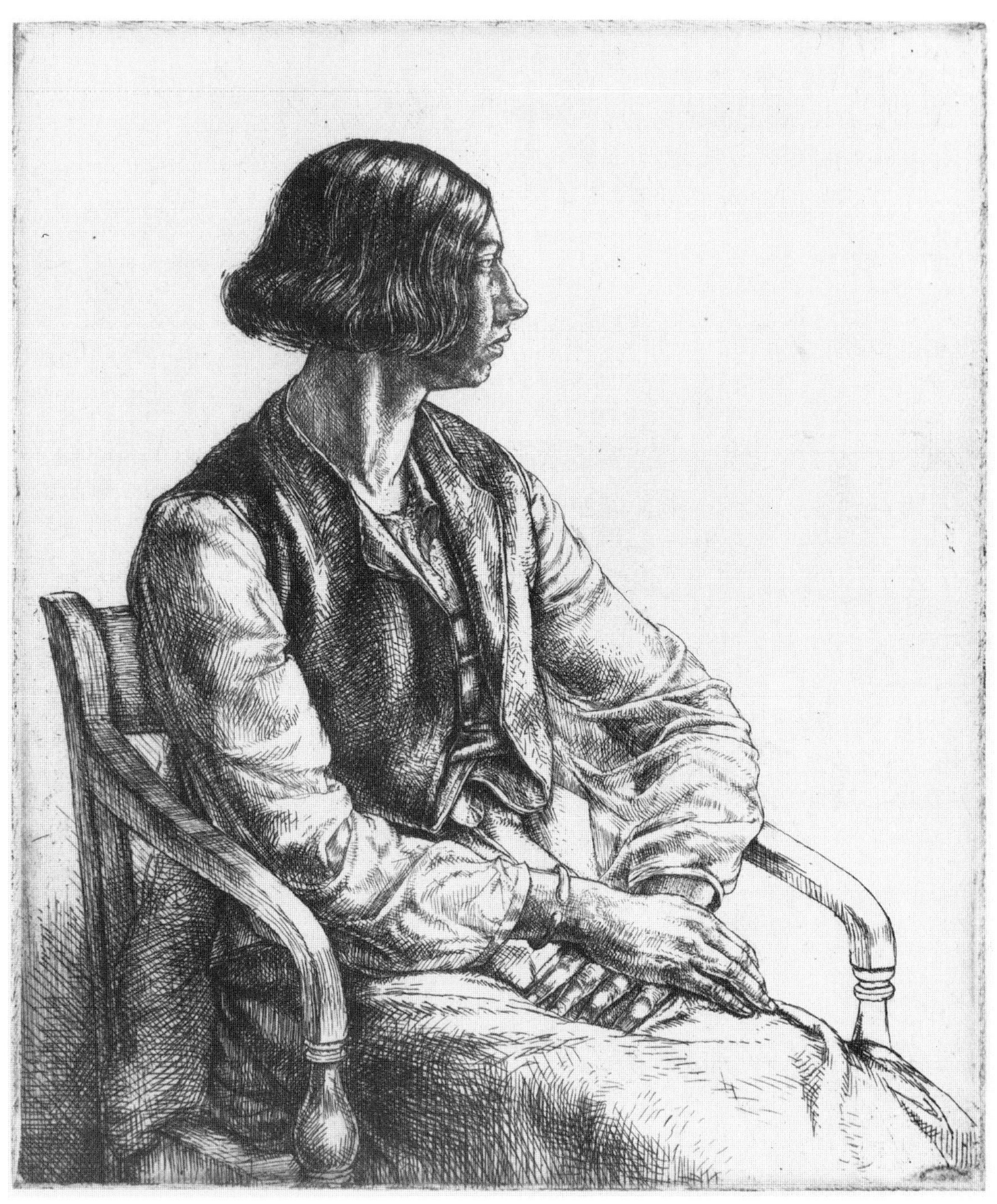

Evelyn Gibbs (1905–1991)   RE344 *Portrait of a Girl* 1928   Etching 180 x 152 mm

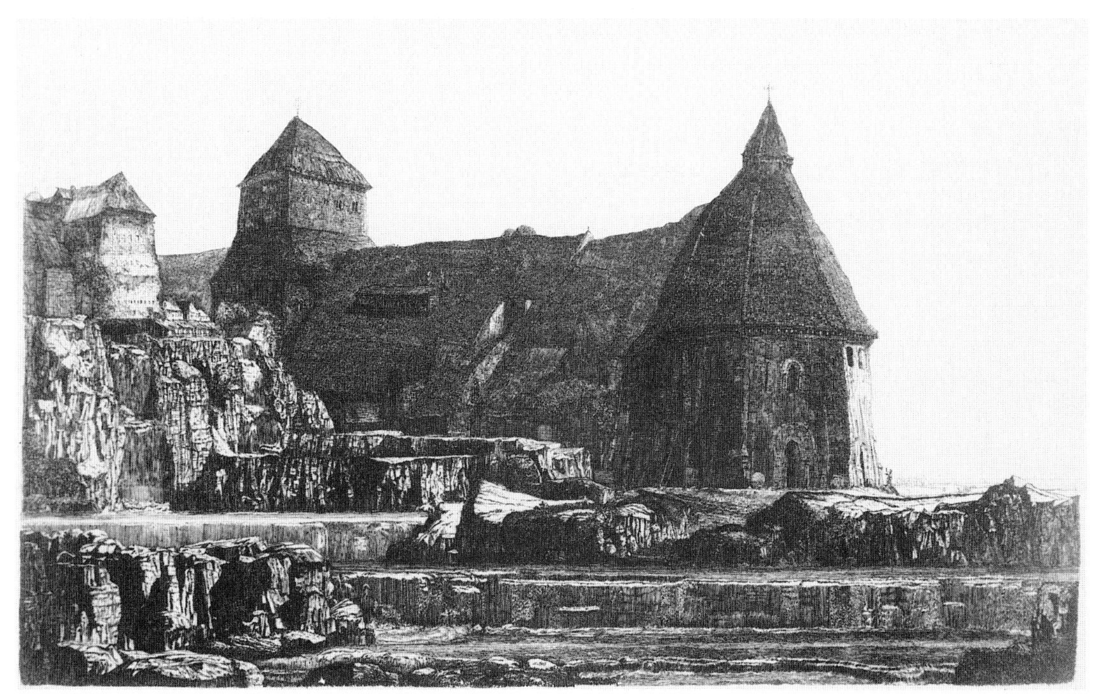

Joseph Webb (1908-1962)   RE348 *The Rat Barn* 1929   Etching 175 x 300 mm

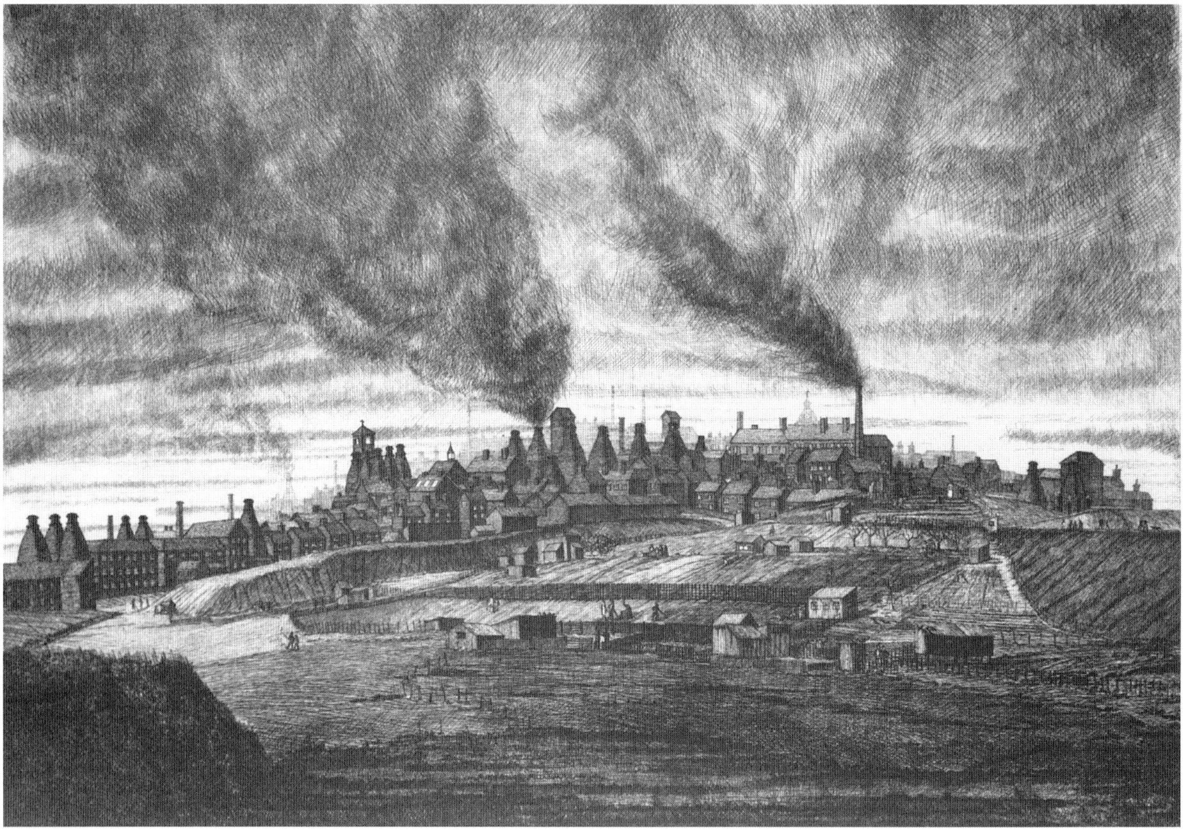

Leonard Griffith Brammer (1906-1994)   RE357 *Burslem* 1930   Etching 243 x 360 mm

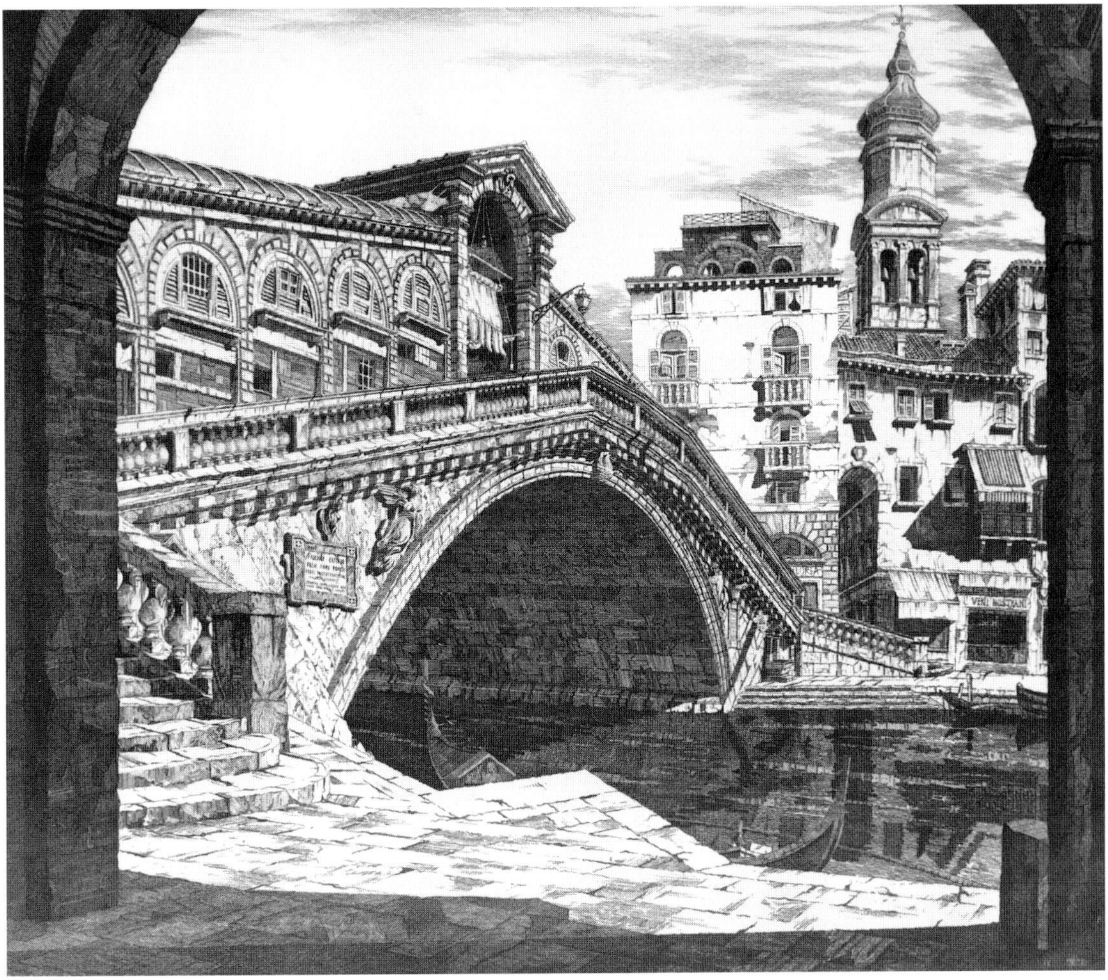

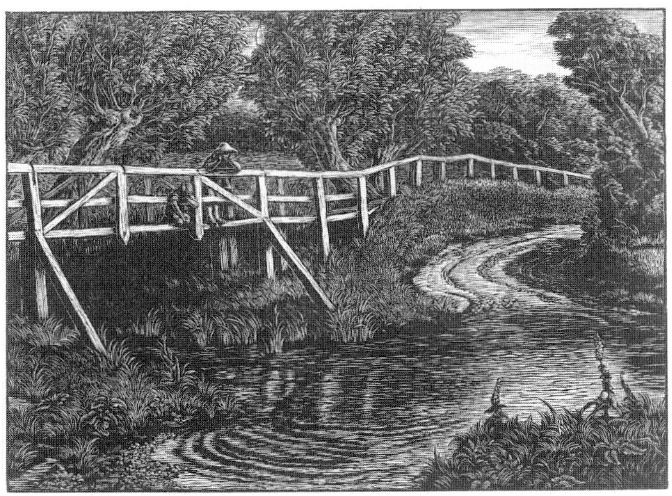

John Taylor Arms (1887–1953)   RE360 *Shadows of Venice* 1930   Etching 256 x 302 mm

Joan Hassall (1906–1988)   RE383 *The Water Splash*   1937
Wood engraving 76 x 108 mm

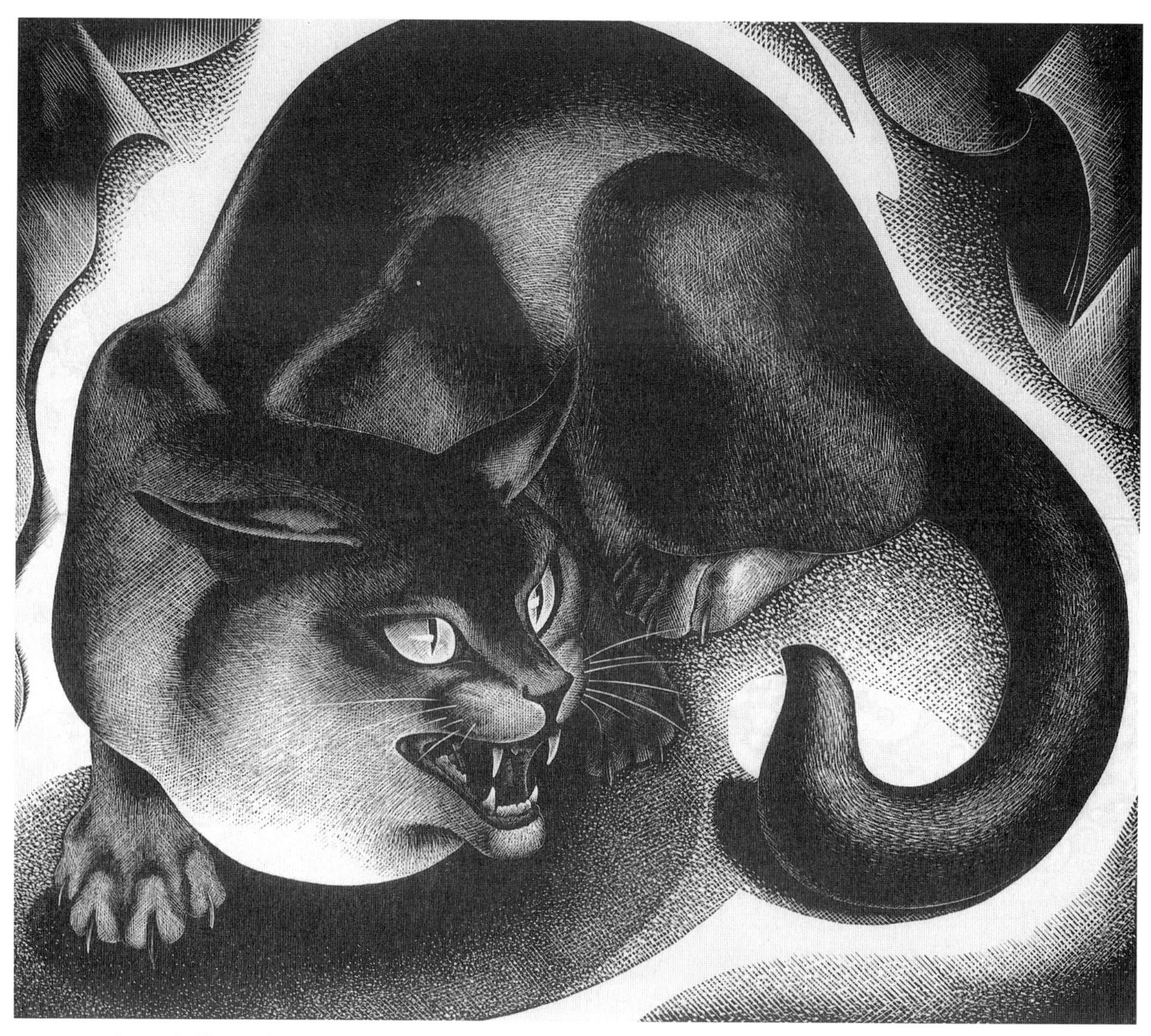

Agnes Miller Parker (1895–1980)   RE386 *The Challenge* 1934   Wood engraving 142 x 163 mm

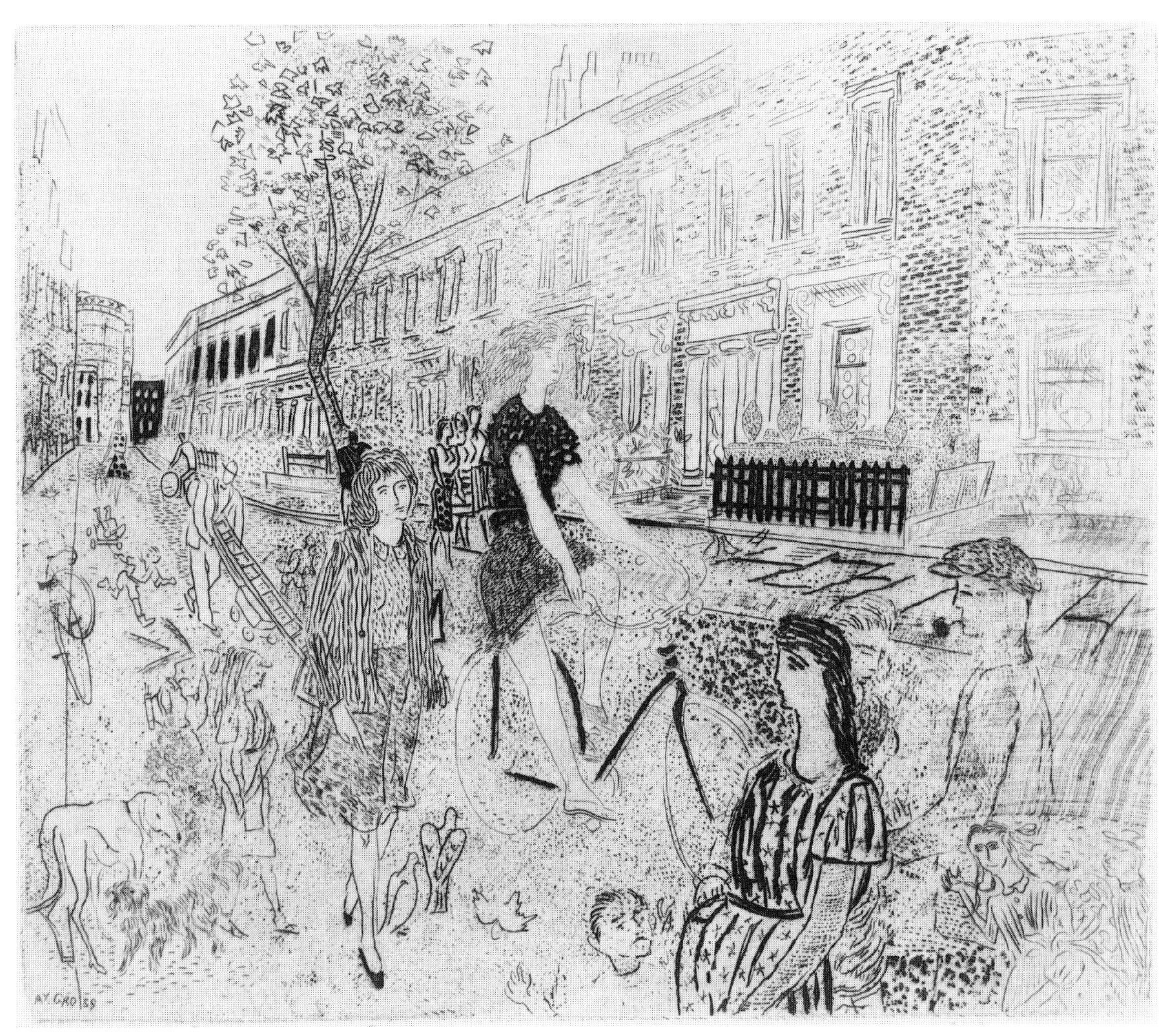

Anthony Gross, C.B.E., R.A., A.R.W.S. (1905-1984)   RE407 *Girl on Bicycle, Fulham* 1947   Etching 170 x 200 mm

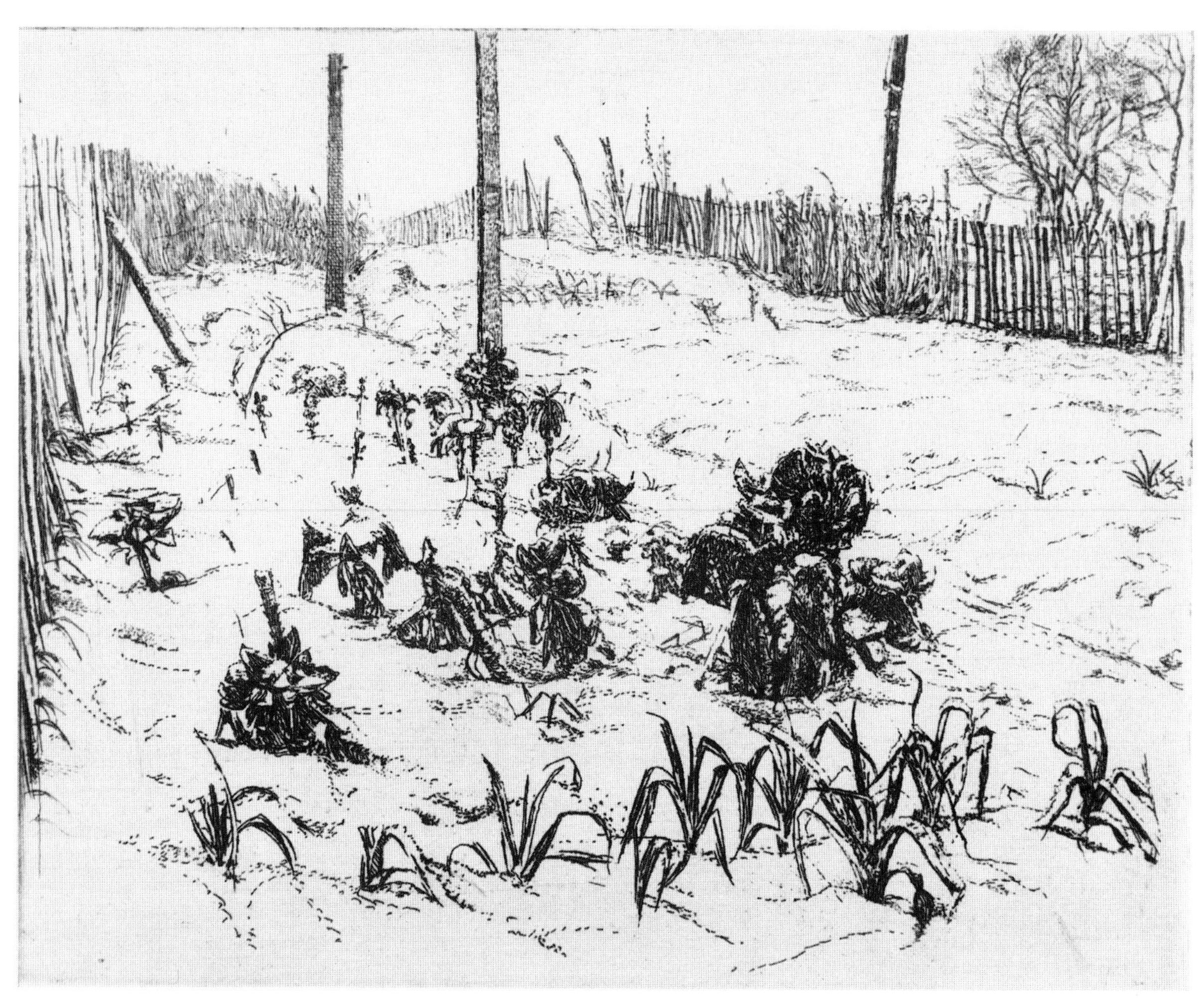

Harry Norman Eccleston, O.B.E., P.P.R.E., R.W.S. (b.1923)   RE418 *The Cabbage Patch* 1948
Etching 128 x 160 mm

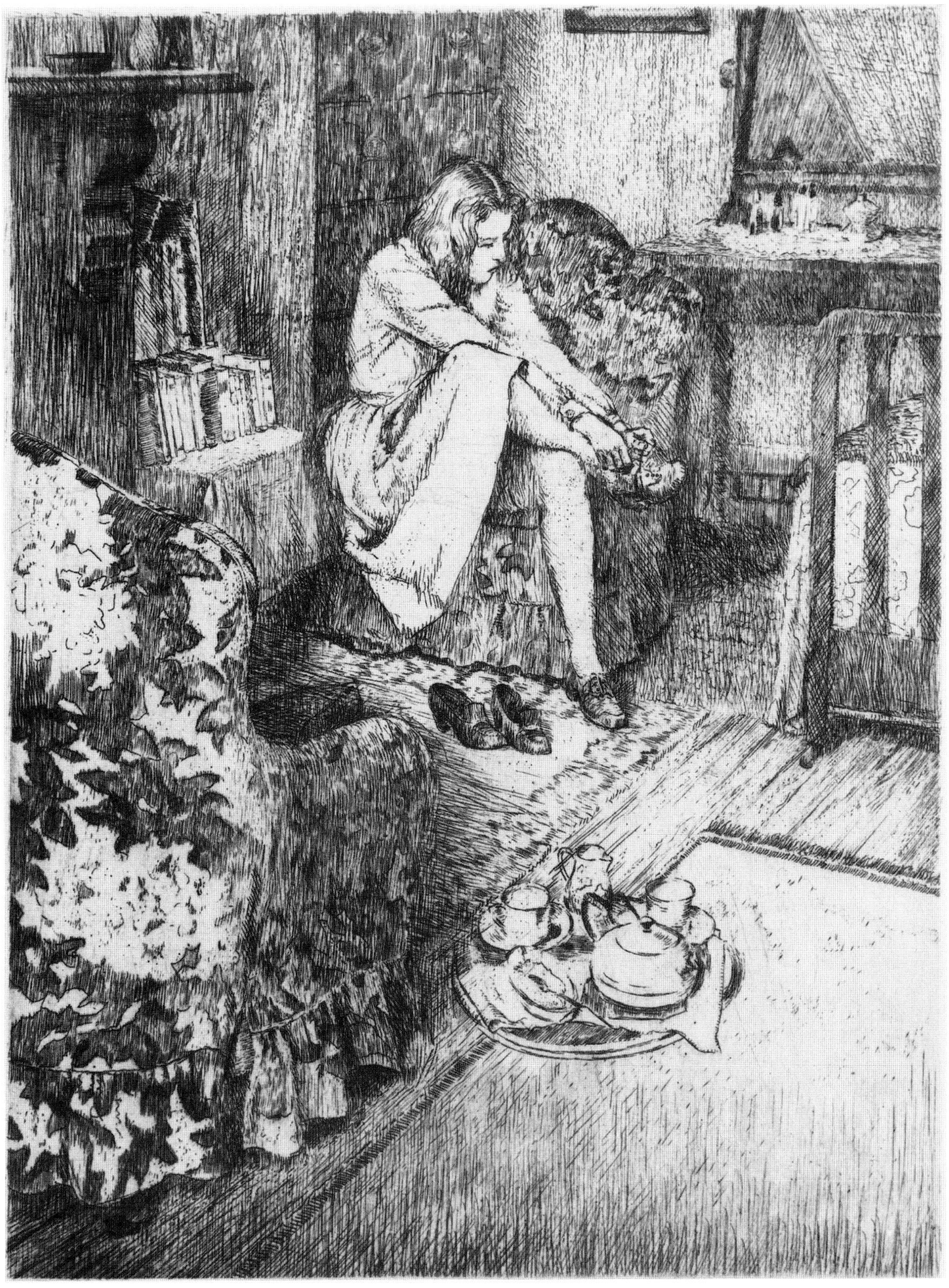

Harold Charles Bartlett, P.P.R.W.S. (b.1921)   RE426 *Girl putting on Shoes*   Etching 202 x 154 mm

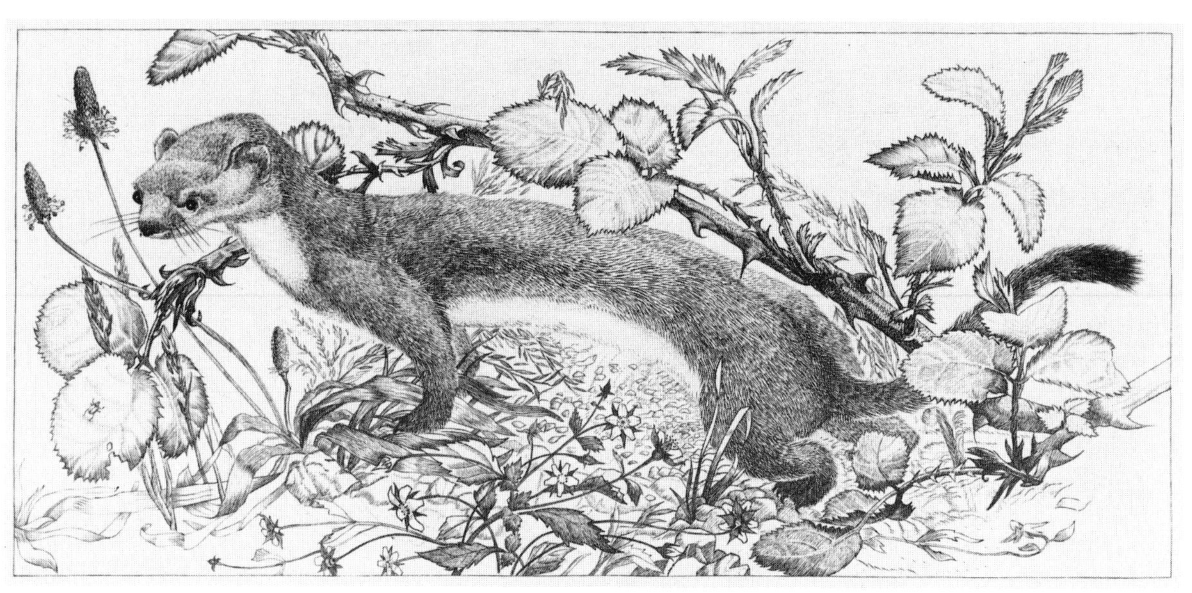

Charles Herbert Arthur Chaplin (1907-1986/7)   RE427 *Stoat* 1950   Engraving 92 x 201 mm

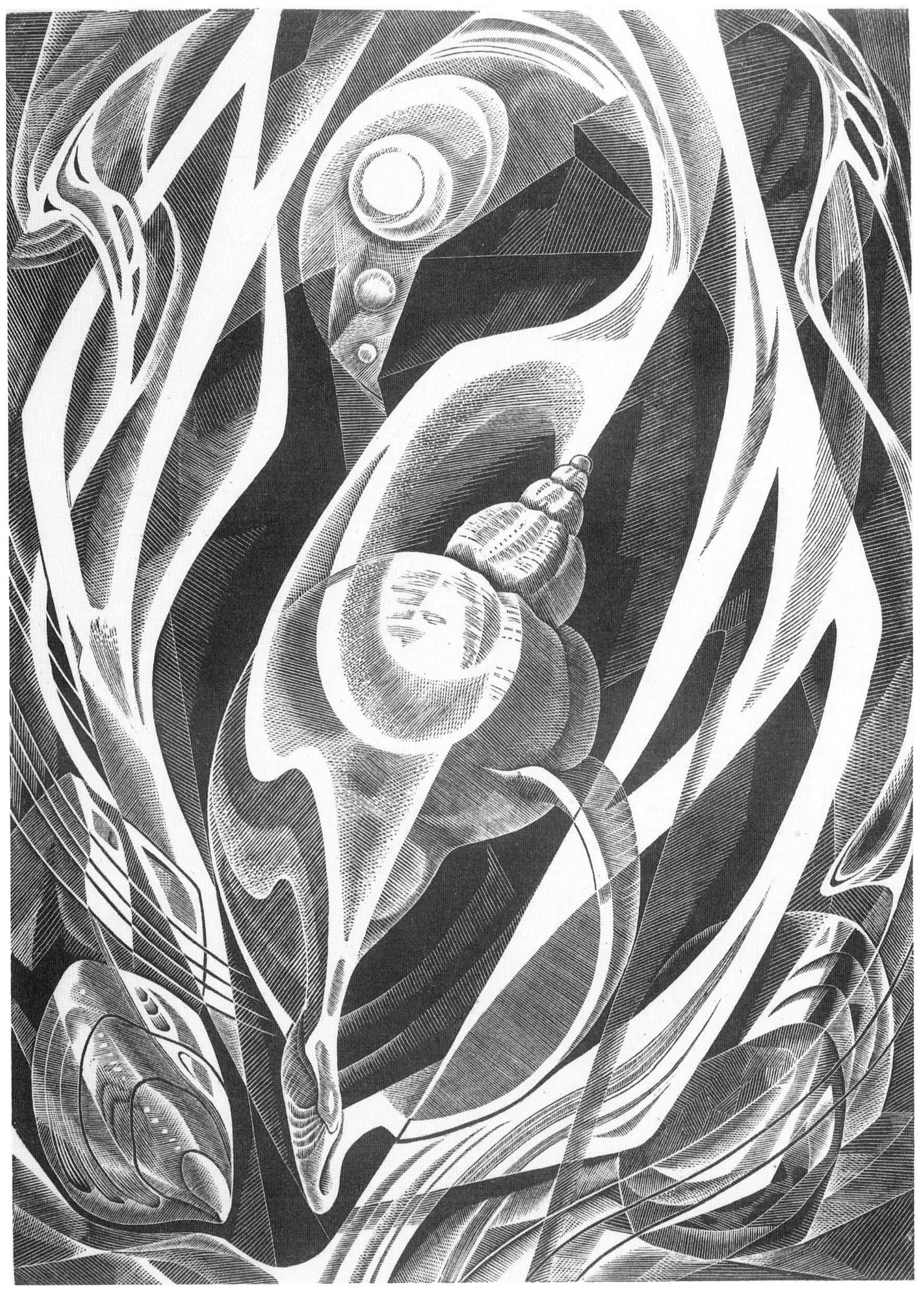

Monica Poole (b.1921)   RE523   *Shell*   1967   Wood engraving 230 x 168 mm

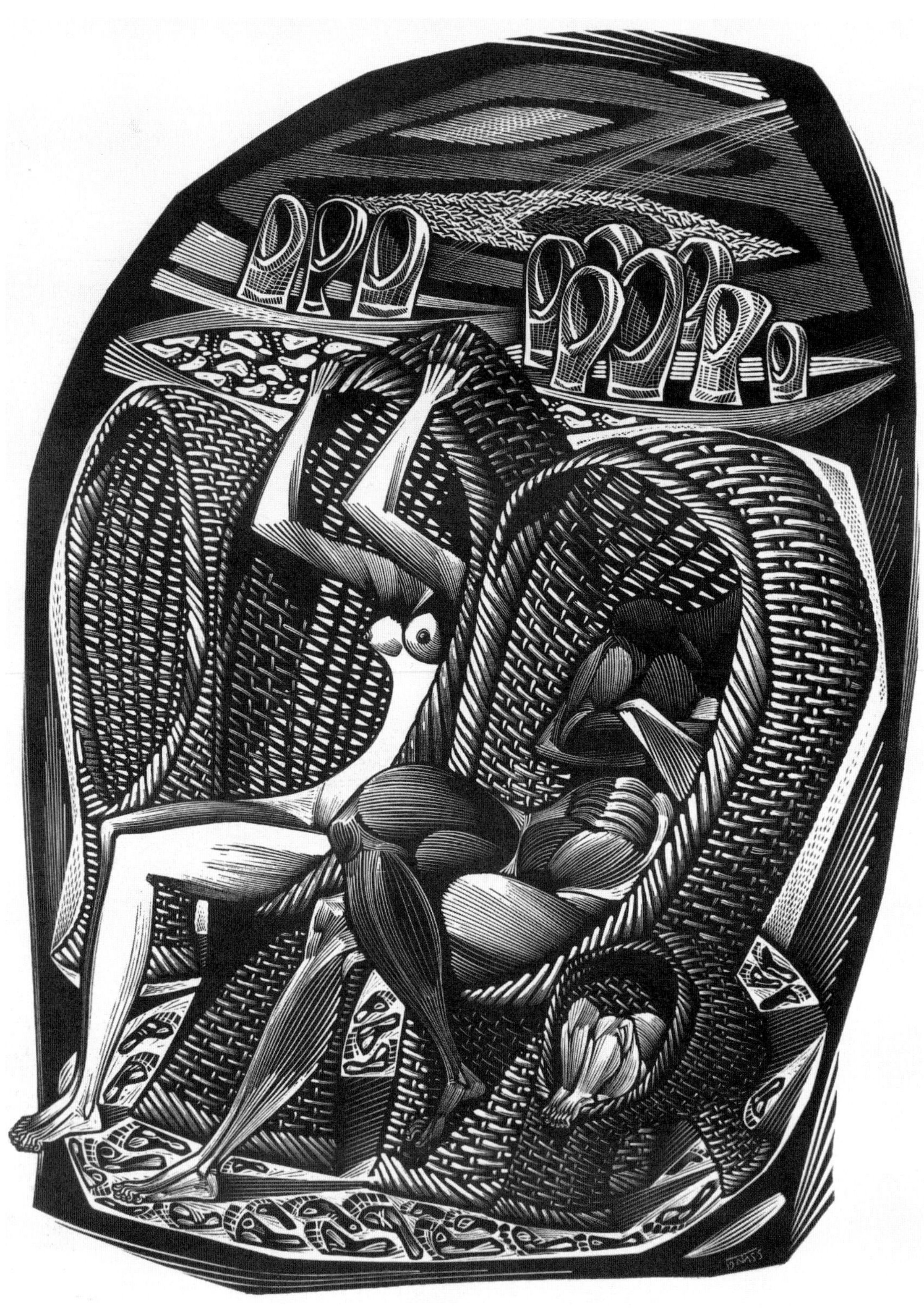

Tranquillo Marangoni (1912–1992)   RE533 *Vimini sulla spiaggia a Scheveningen*   1955
Wood engraving 315 x 230 mm

112

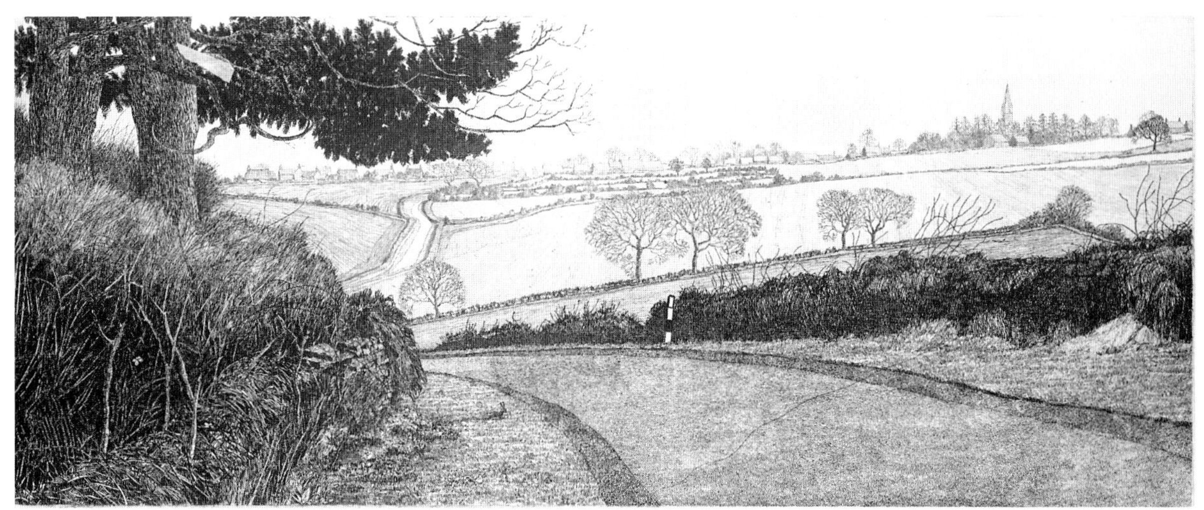

Joseph William Winkelman, P.P.R.E., Hon. R.W.S. (b.1941)   RE585 *Hatching Lane, Leafield* 1979   Etching 190 x 467 mm

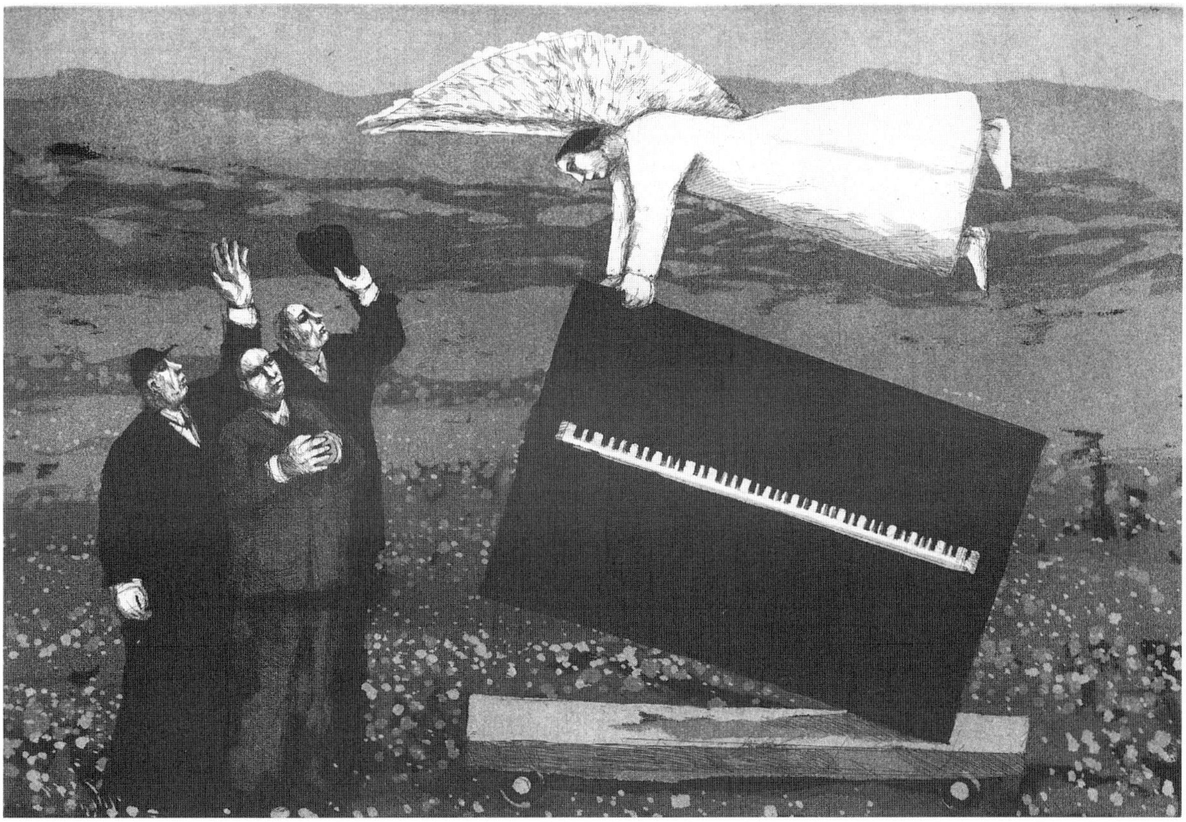

Karólína Lárusdóttir, R.W.S. (b.1944)   RE603 *The Piano Movers Thank the Angel*
Etching & aquatint 195 x 292 mm

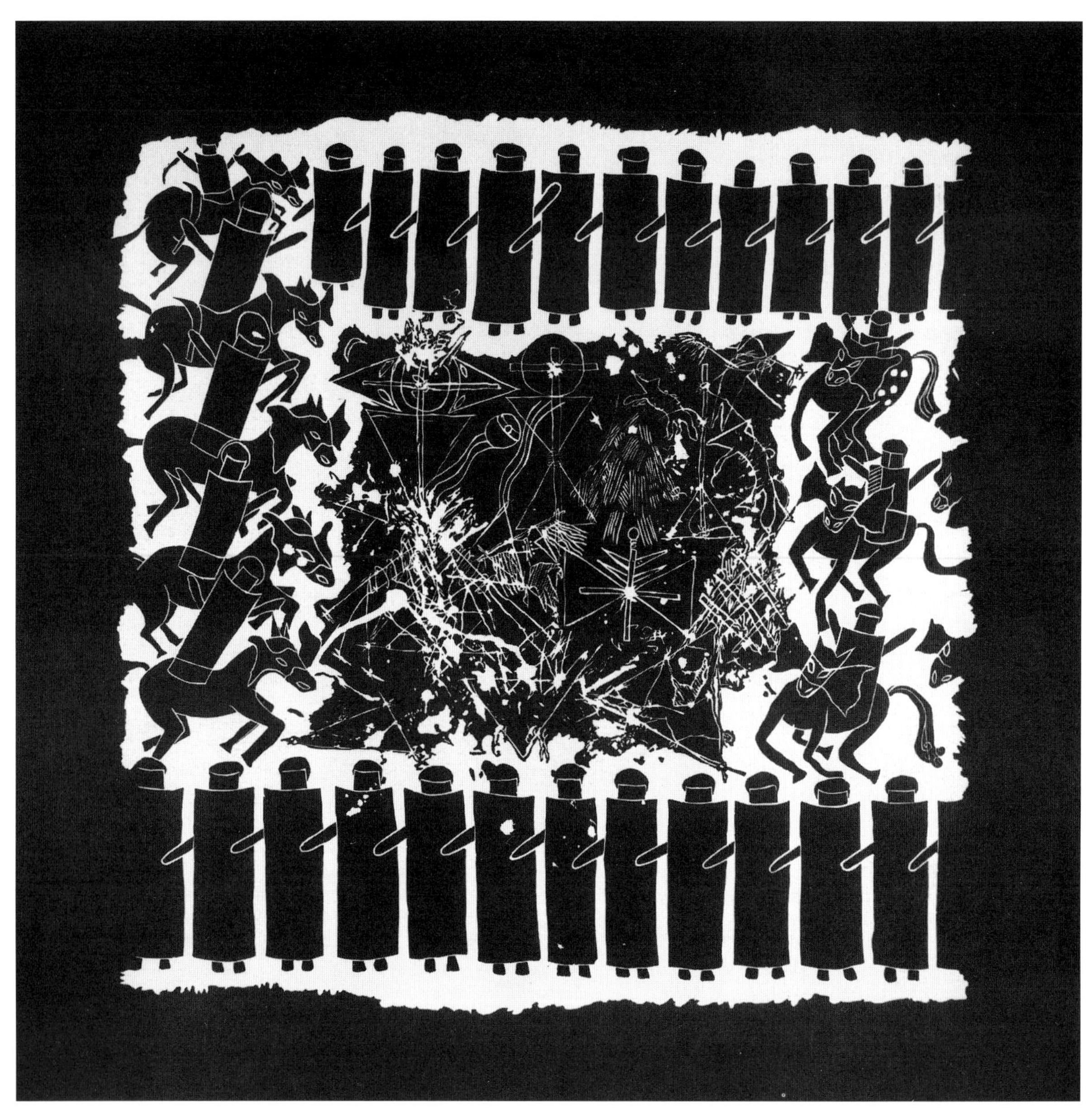

Sandy Sykes (b.1944)   RE683 *Contained*   Linocut 650 x 663 mm

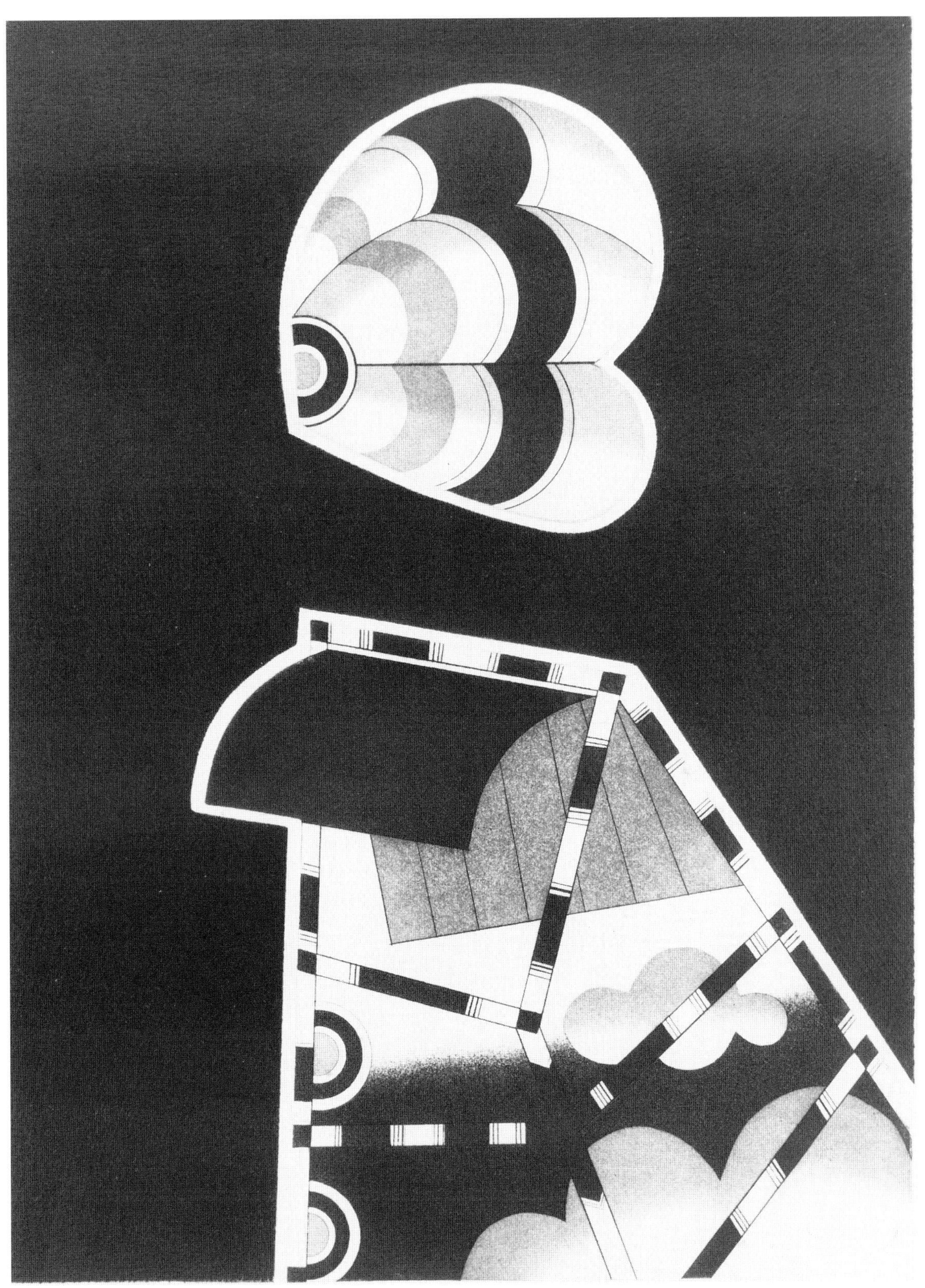

Ted Atkinson, R.B.S. (b.1929)   RE693 *Analogue* 1985   Aquatint 340 x 251 mm

115

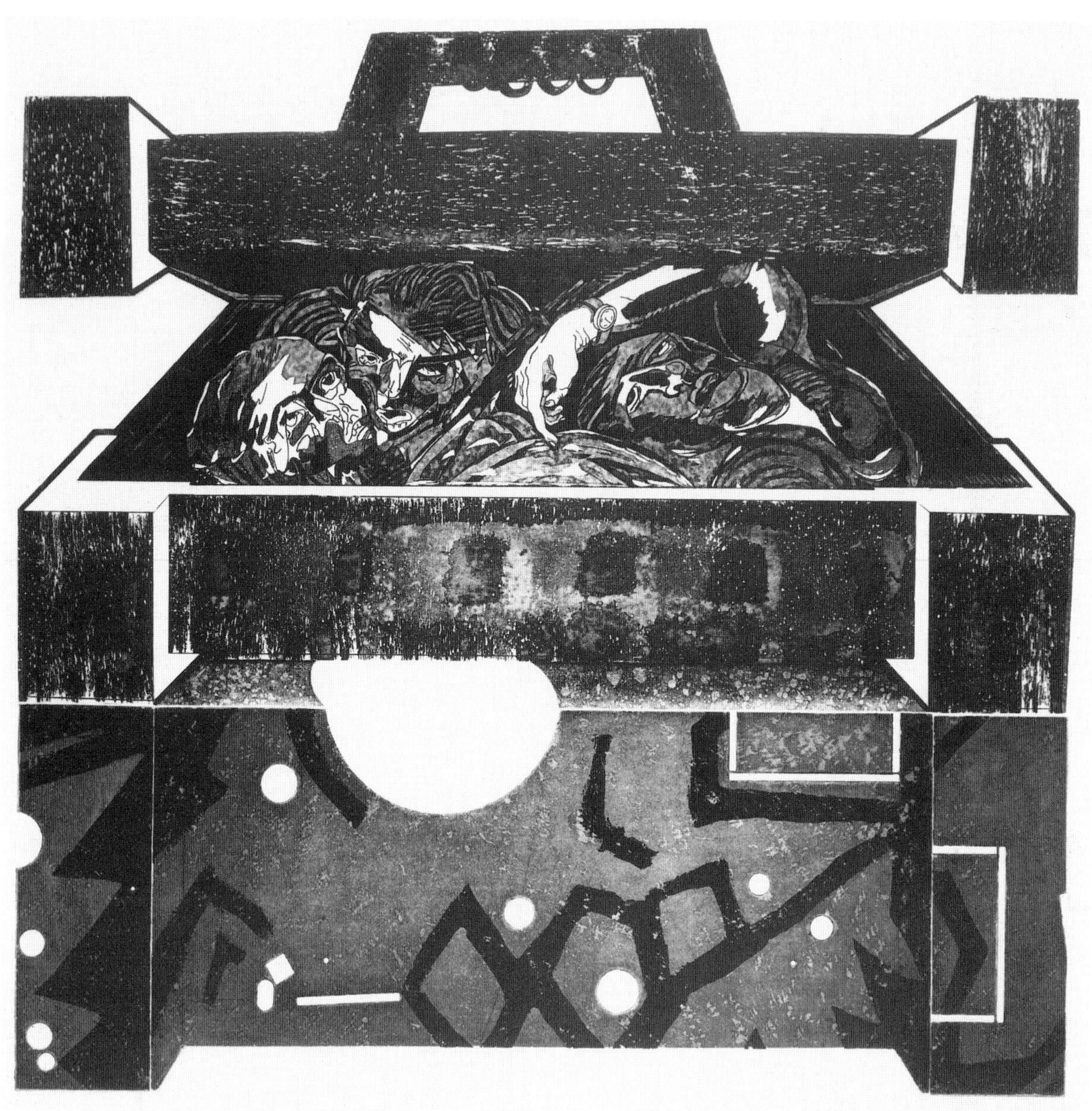

Ann Westley (b.1948)   RE732 *Black Box Throws Light*   Etching & aquatint 545 x 553 mm

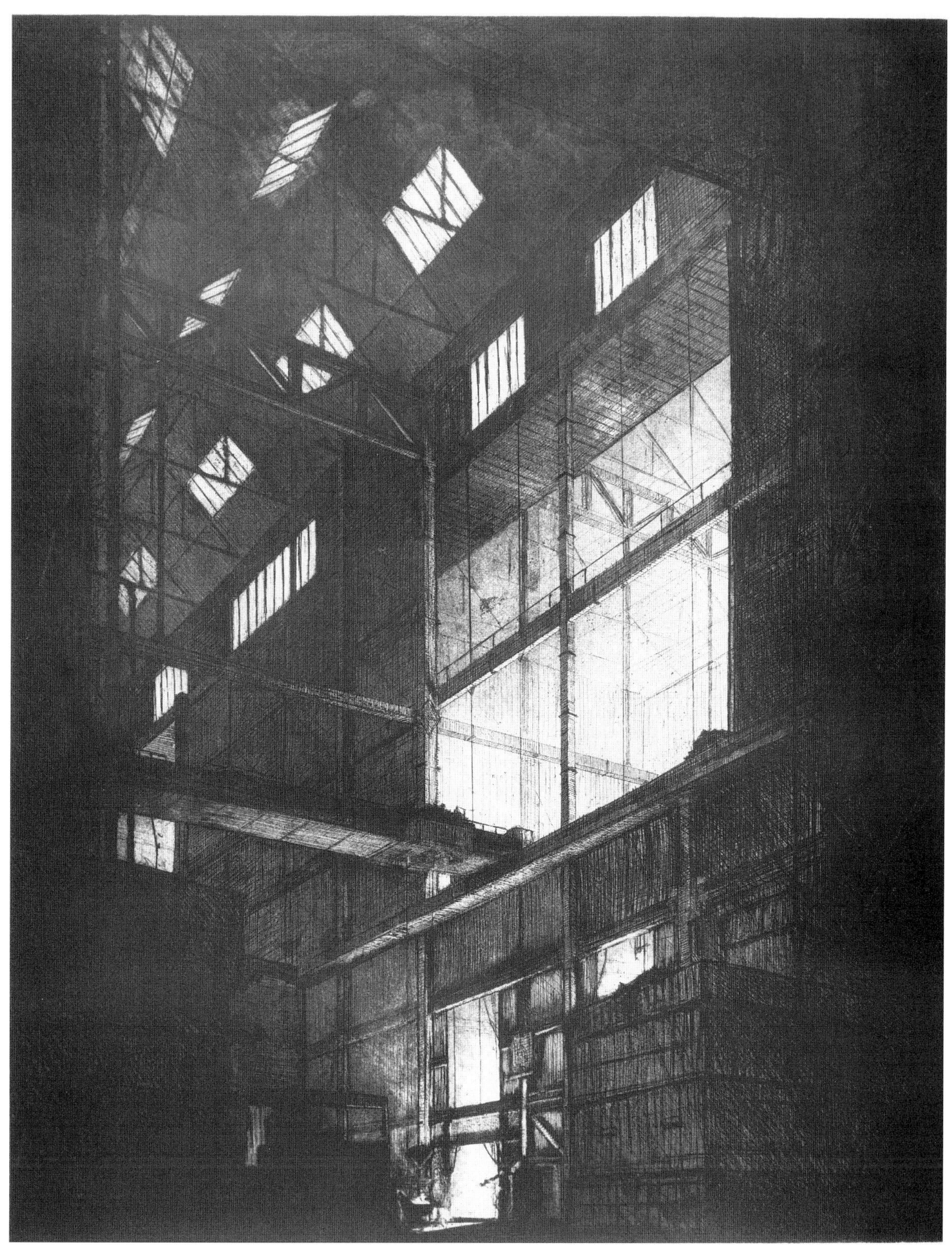

John Howard (b.1958)   RE733 *The Foundry I*   Etching & aquatint 561 x 443 mm

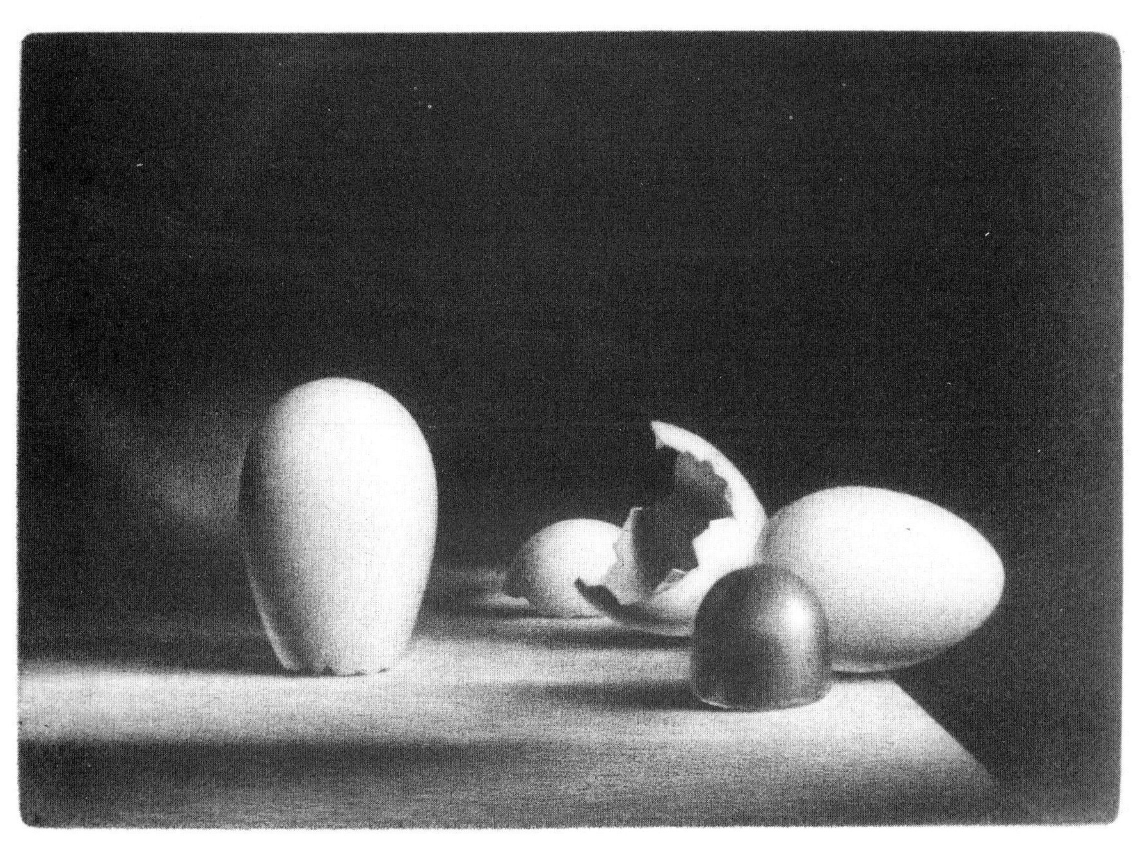

Konstantyn Chmutin (b.1953)   RE746 *"Conception No.6"* 1990   Mezzotint 106 x 152 mm

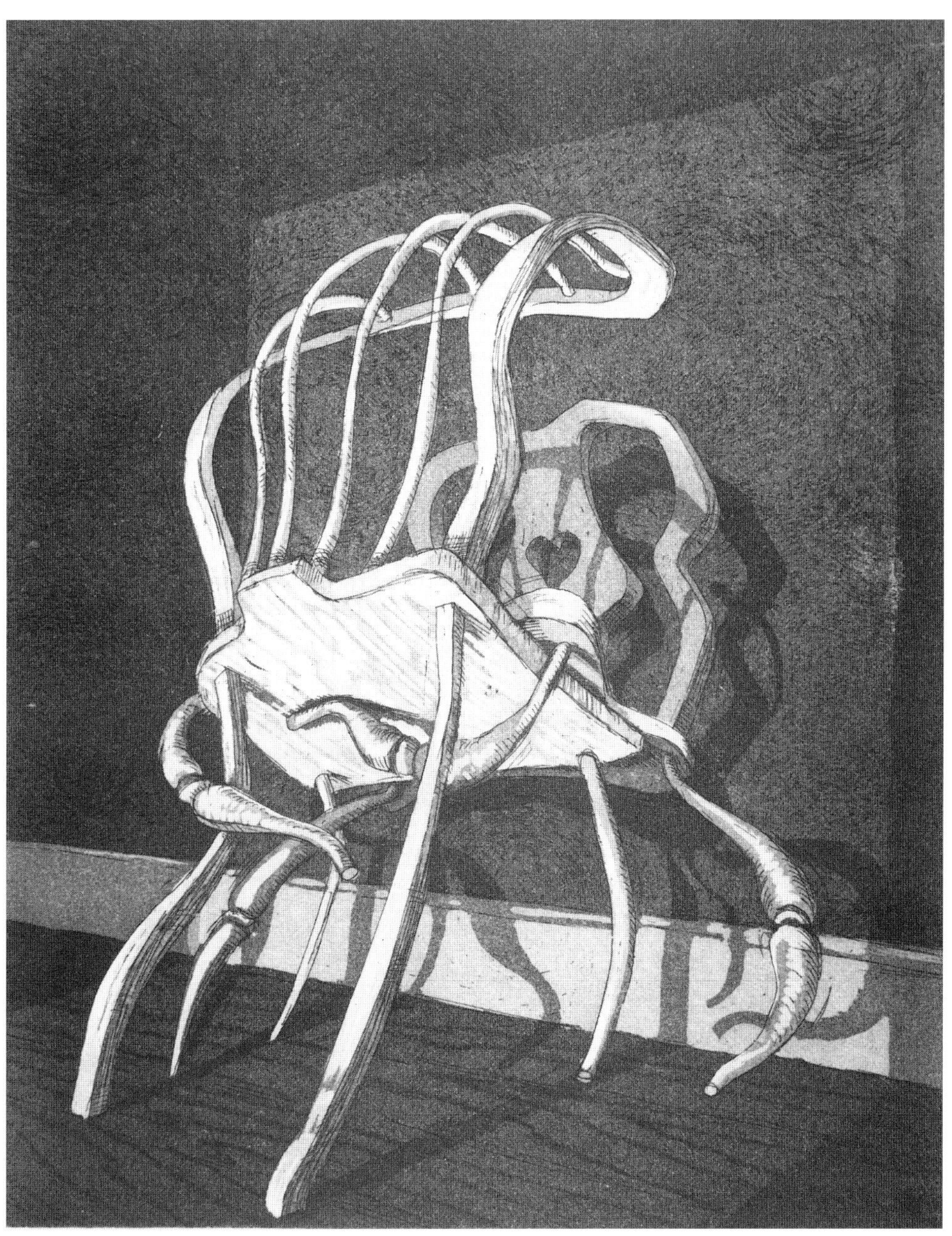

Magnus Irvin (b.1952)   RE770 *After Dark*   Etching & aquatint 165 x 132 mm

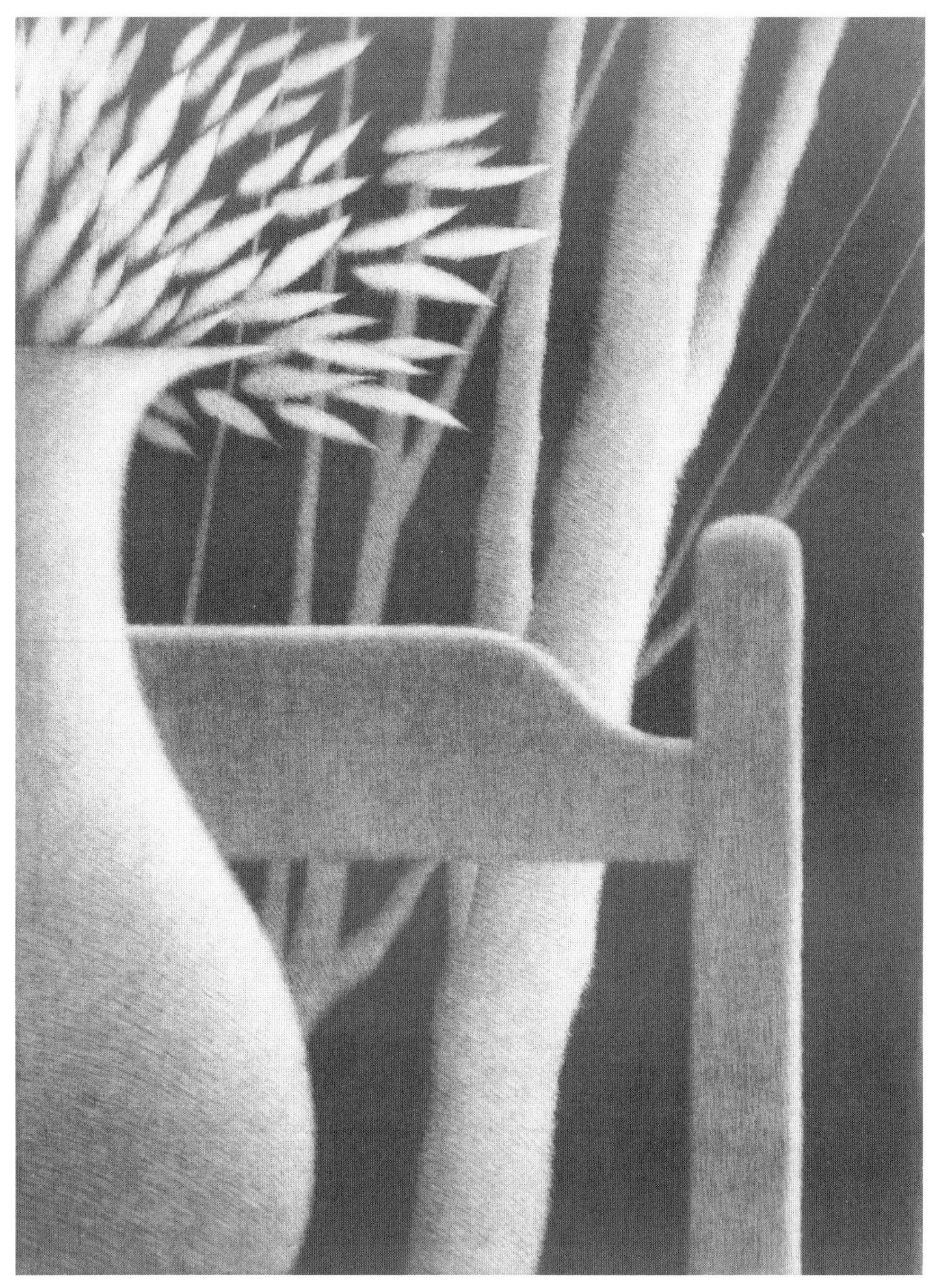

Robert Kipniss (b.1931)   RE801 *Vase, Chair & Trees* 1995   Mezzotint 176 x 137 mm

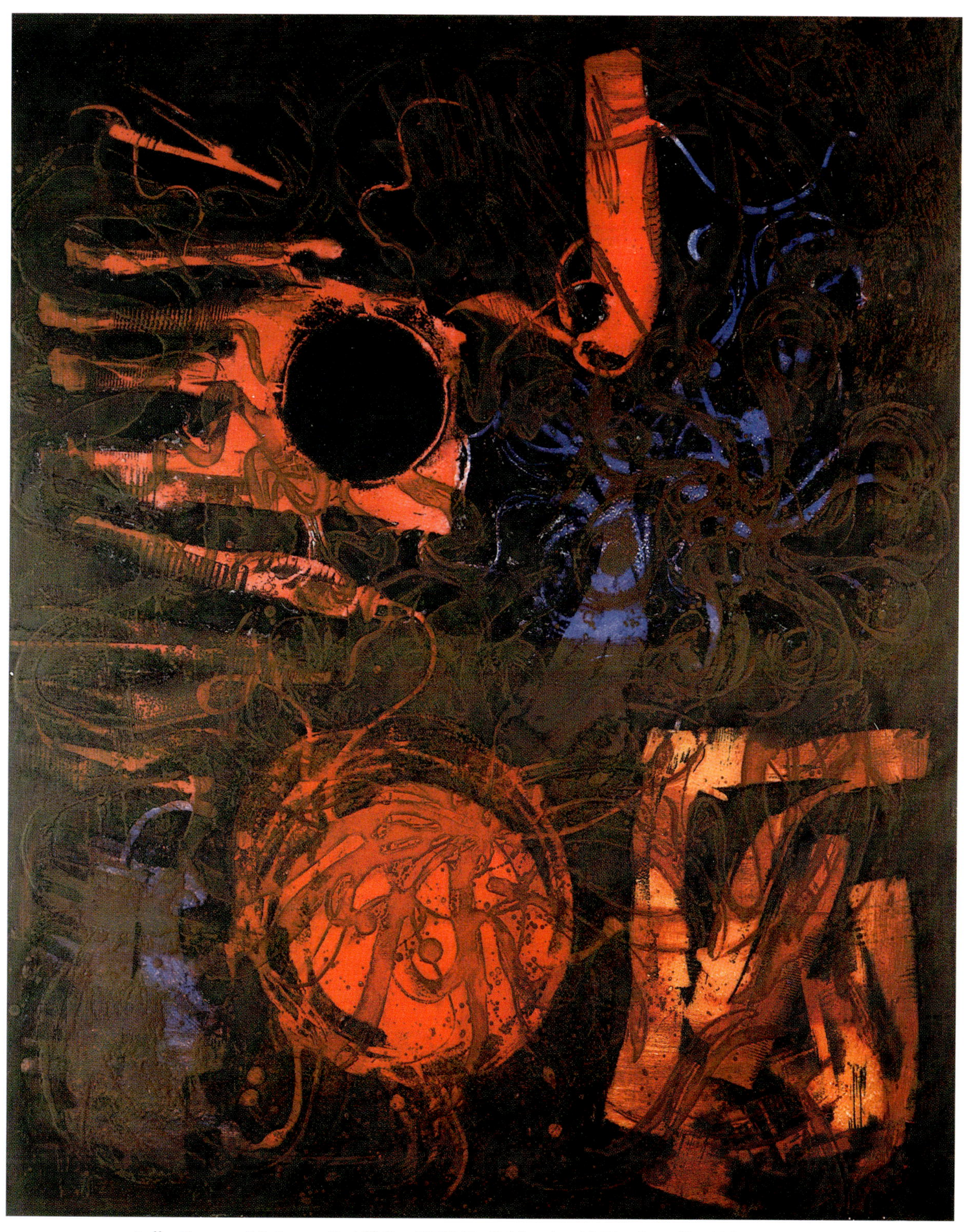

Sally Frances McLaren (b.1936)   RE503 *Celebration* 1961   Intaglio 610 x 500 mm

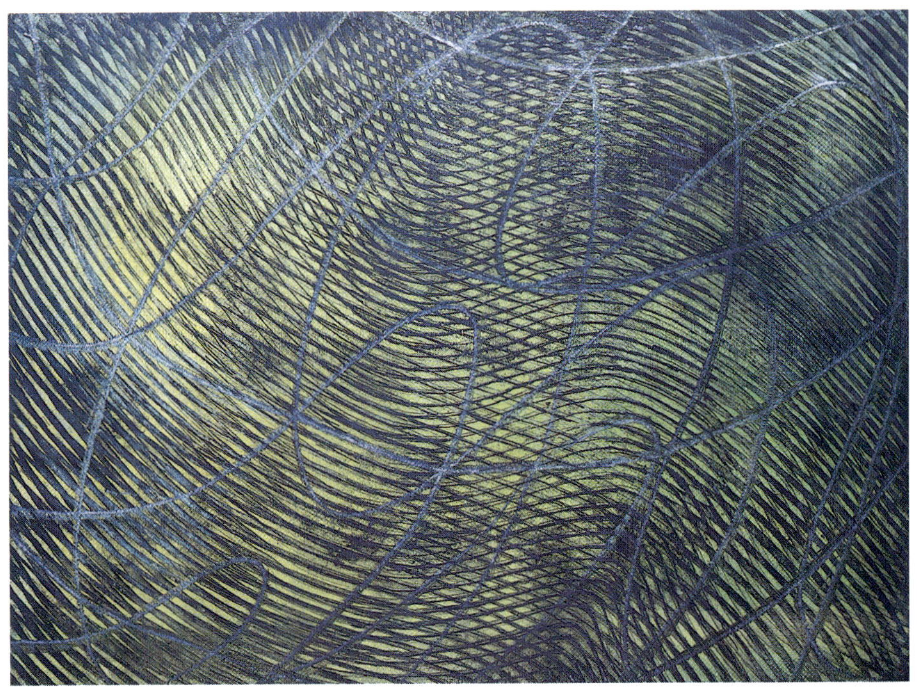

Stanley William Hayter, C.B.E., Hon.R.A. (1901–1988)   RE520 *Lunar Rhythm* 1967
Etching & aquatint 347 x 483 mm

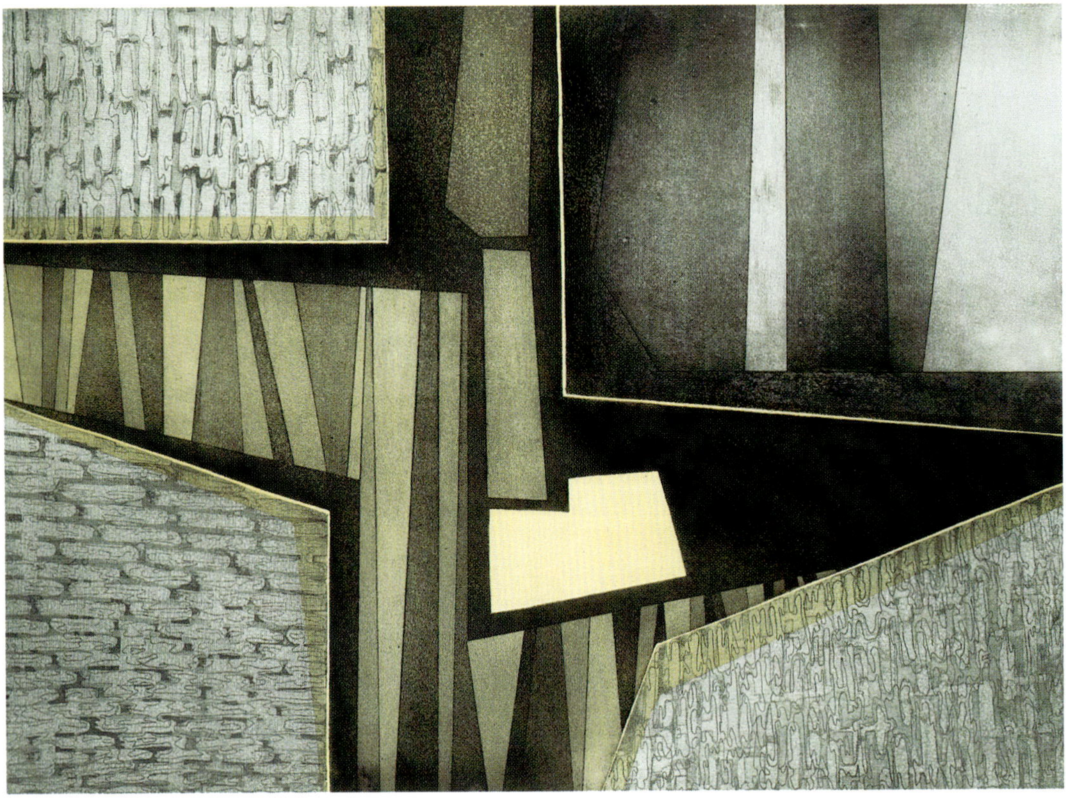

Vikki Slowe (b.1947)   RE582/1 *Pavements* 1978   Etching & aquatint 421 x 590 mm

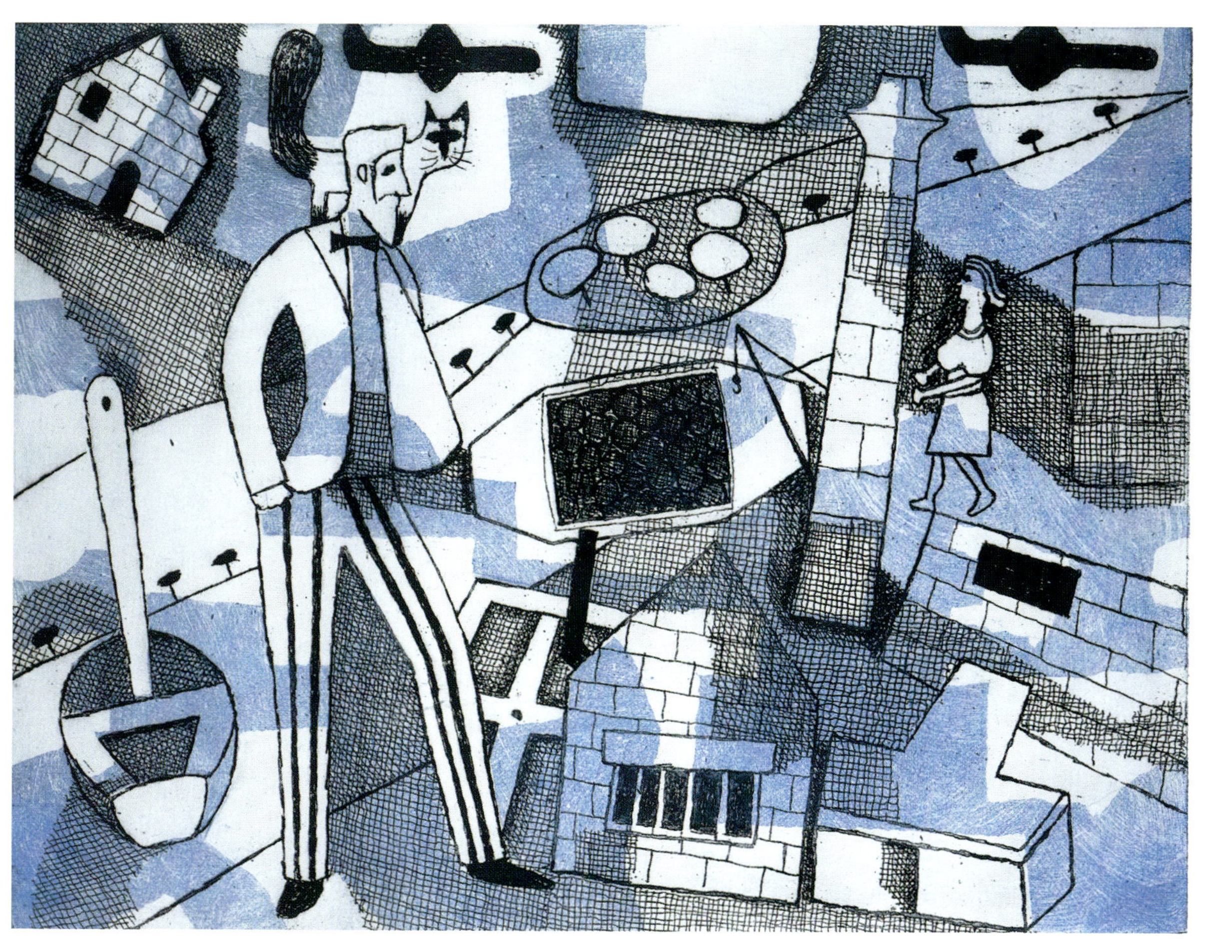

Julian Trevelyan, R.A. (1910–1988)   RE678 *Me and Tom* 1976   Etching & aquatint 352 x 475 mm

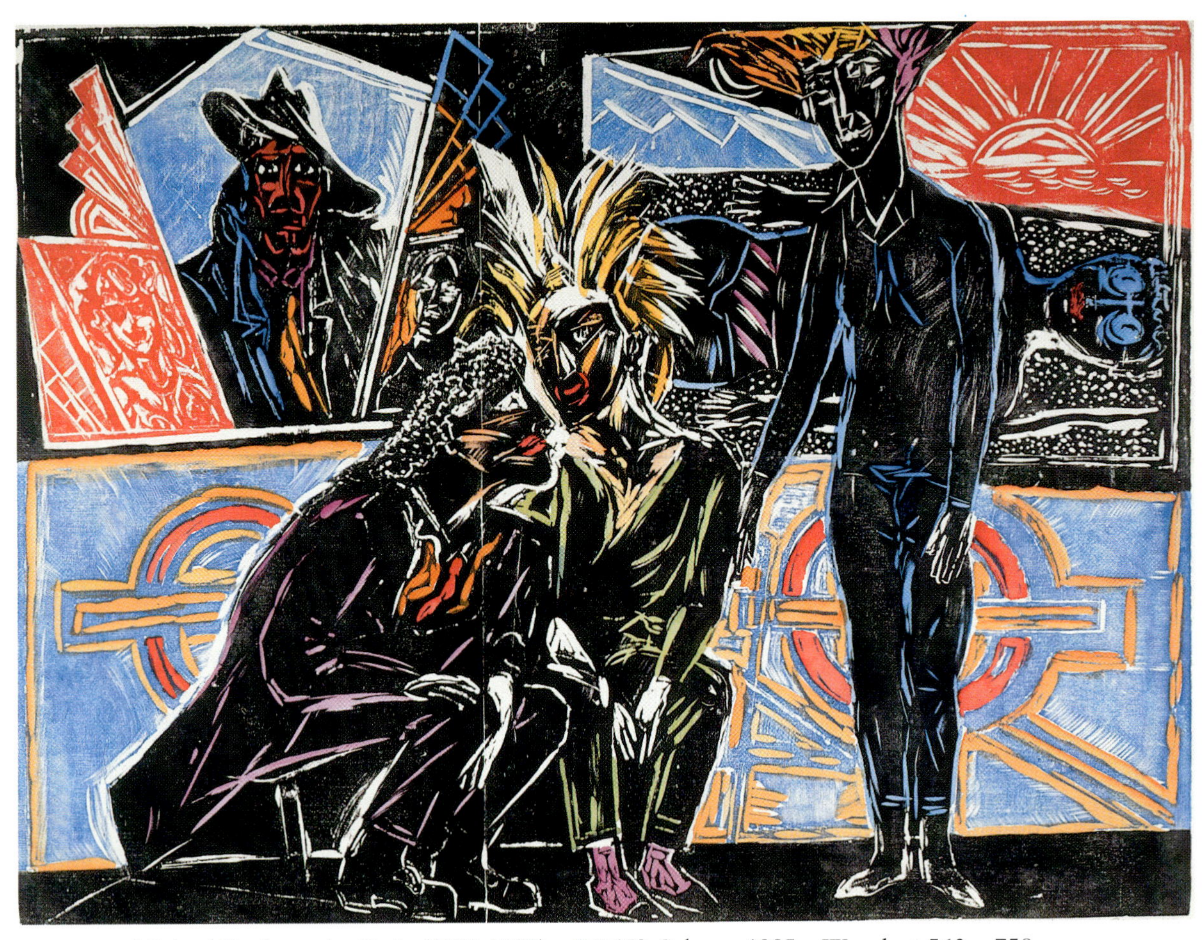

Michael Rothenstein, R.A. (1908–1993)   RE679 *Subway* 1985   Woodcut 563 x 758 mm

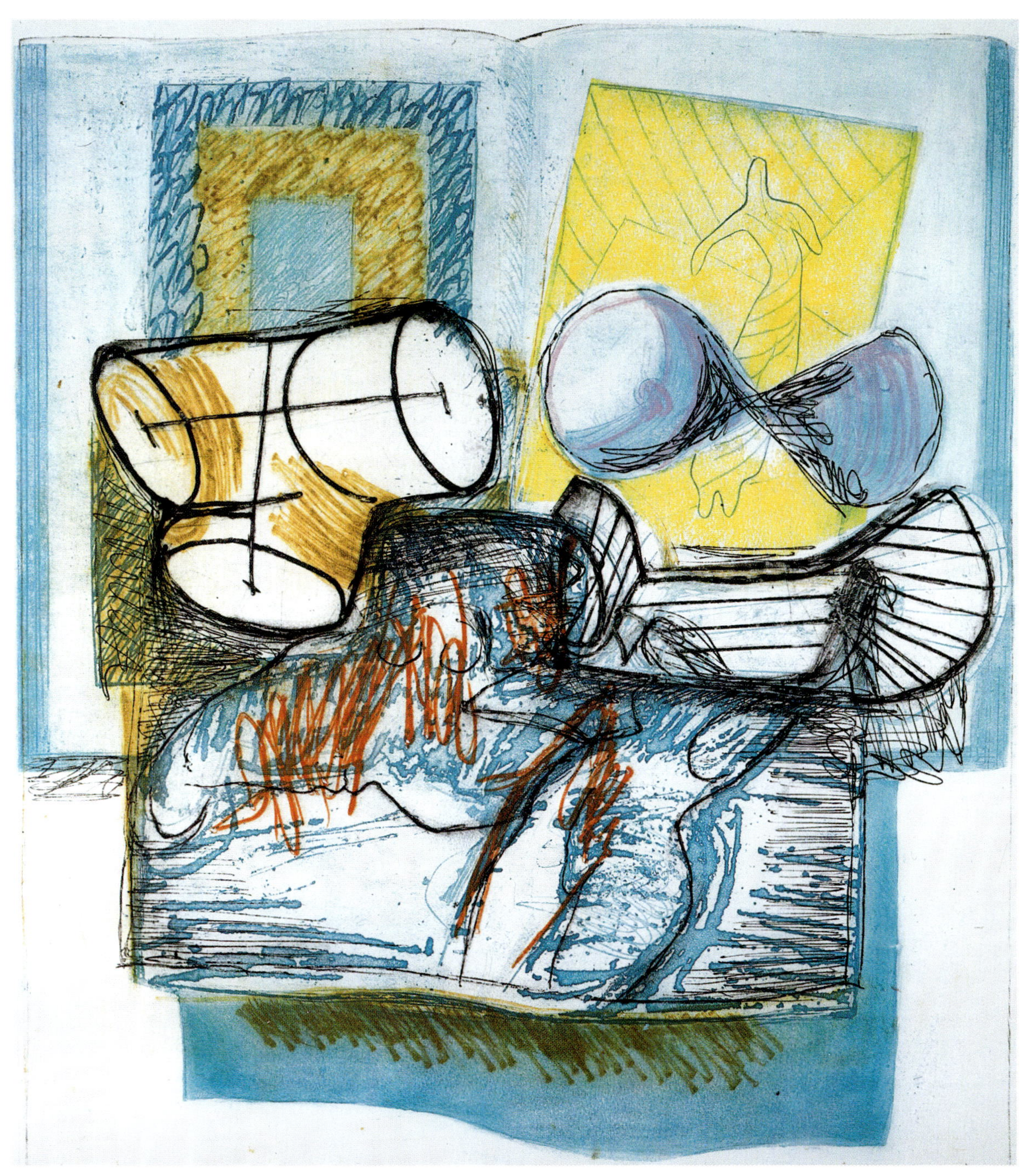

Agatha Sorel, A.R.W.S. (b.1935)   RE681 *Divine Proportions* 1985   Engraving & aquatint 490 x 430 mm

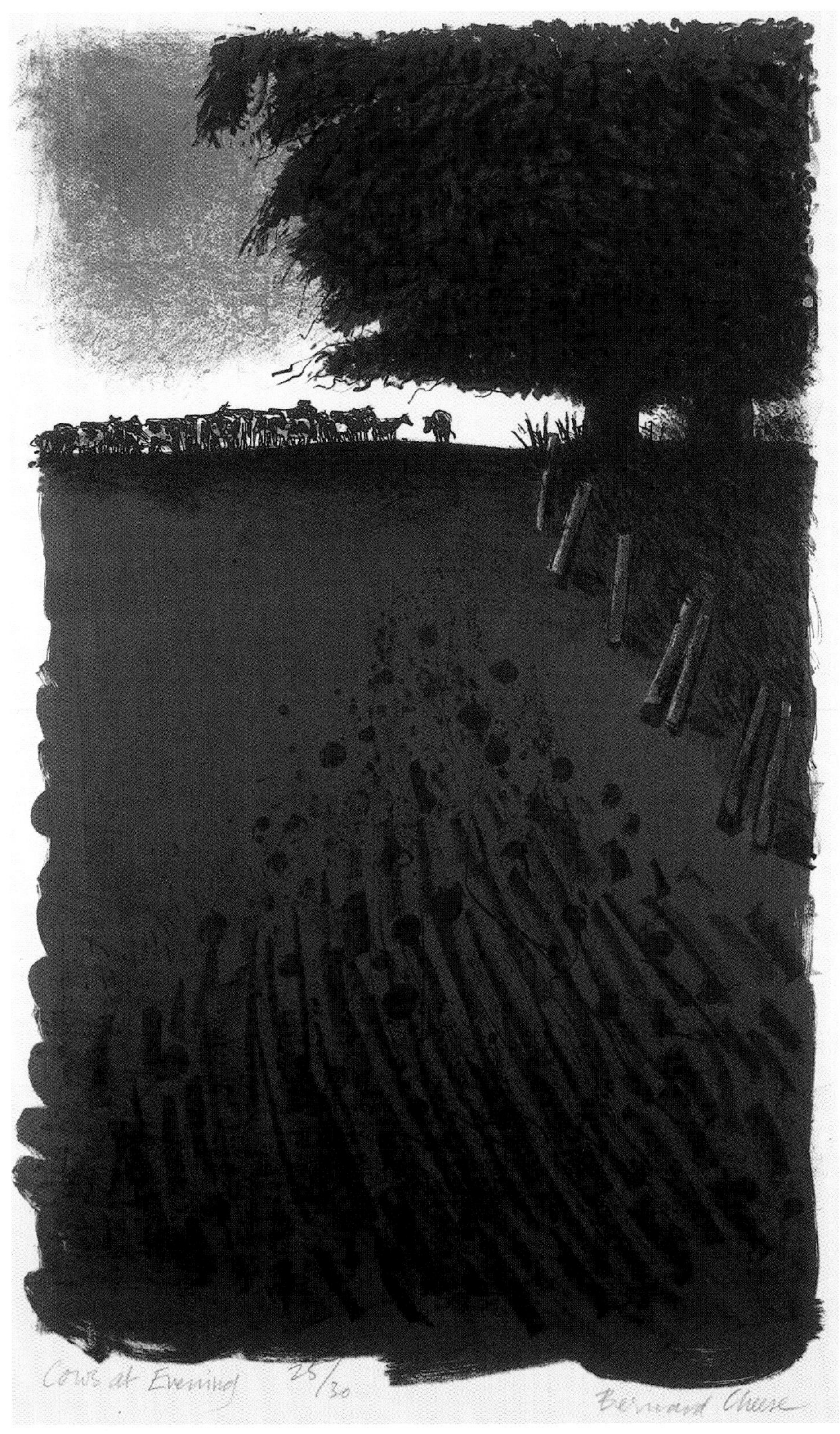

Bernard Cheese (b.1925)   RE706 *Cows at Evening* 1964   Lithograph 465 x 280 mm

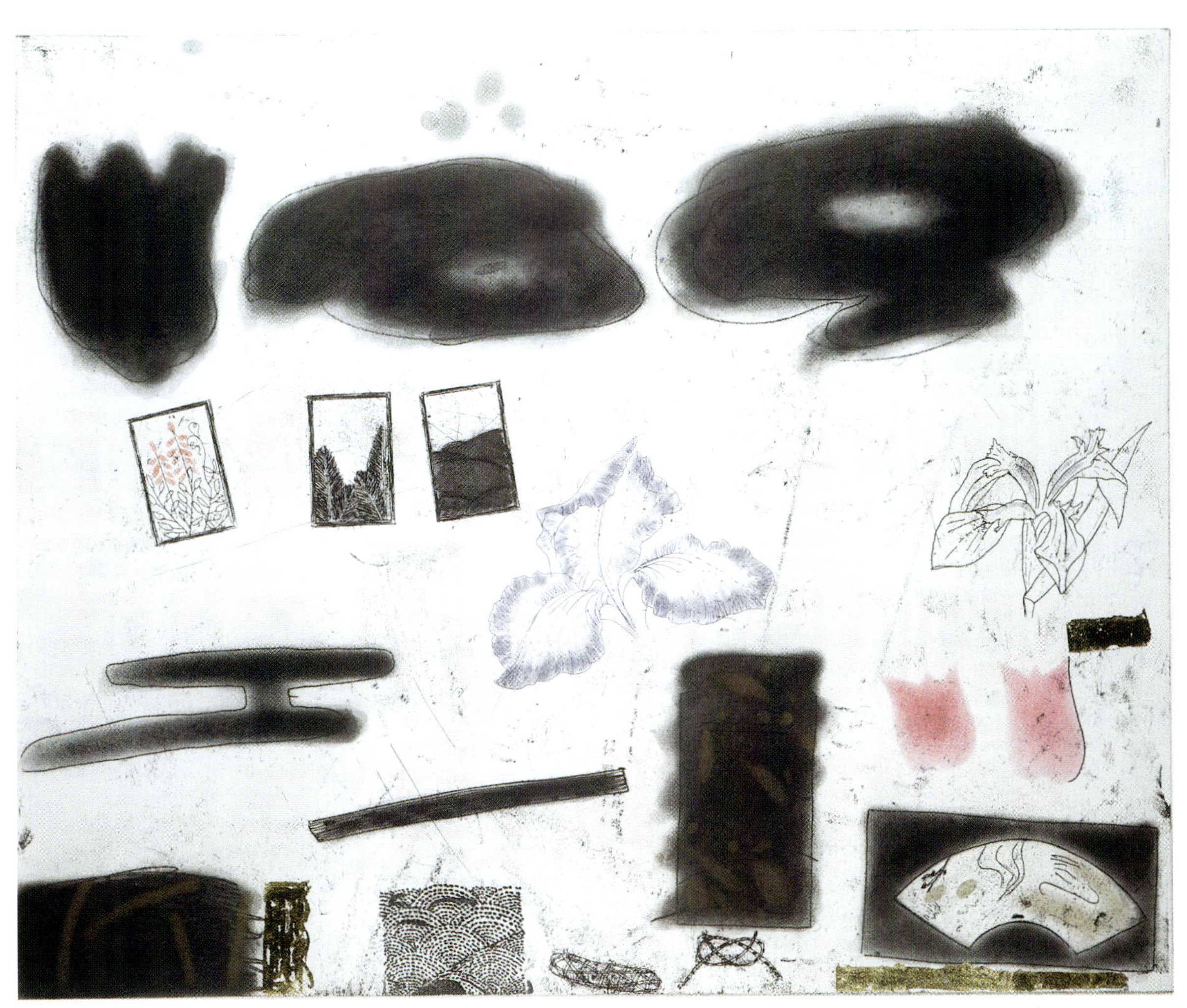

Elizabeth Blackadder, O.B.E., R.A., Hon.R.W.S., R.S.A. (b.1931)   RE773 *Untitled*
Intaglio 430 x 528 mm

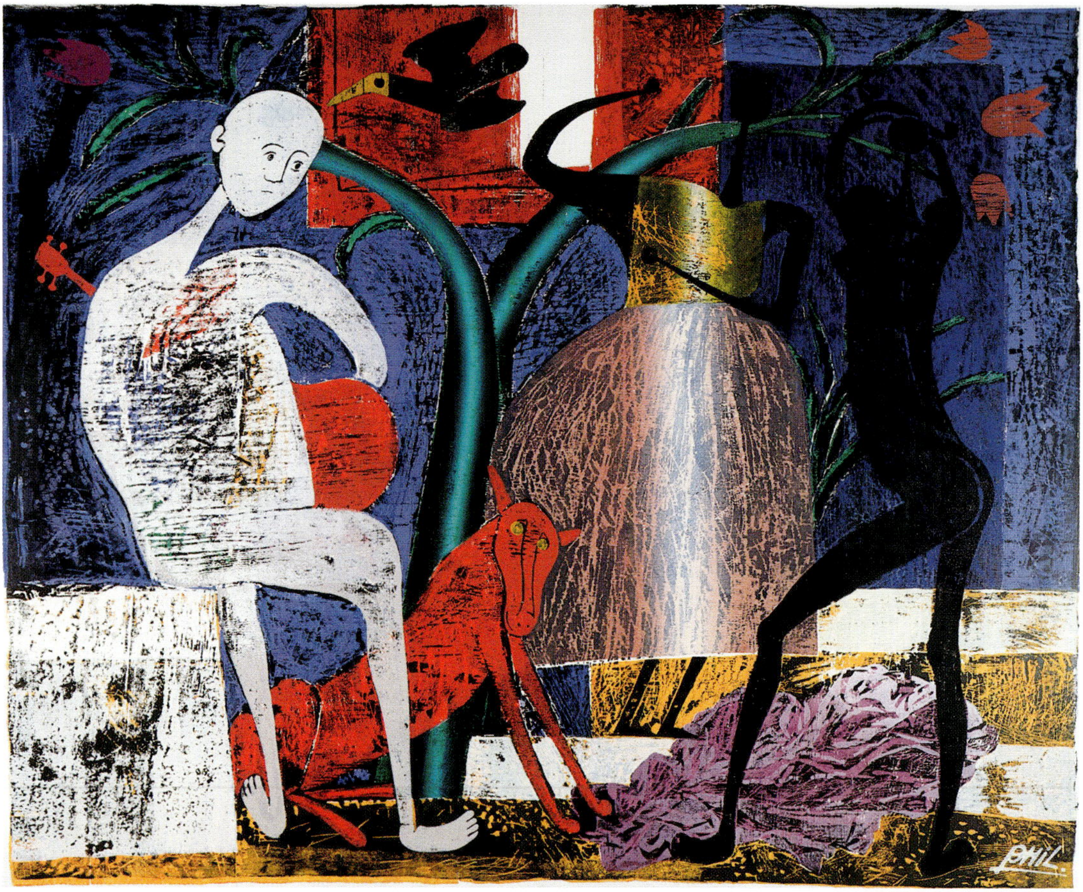

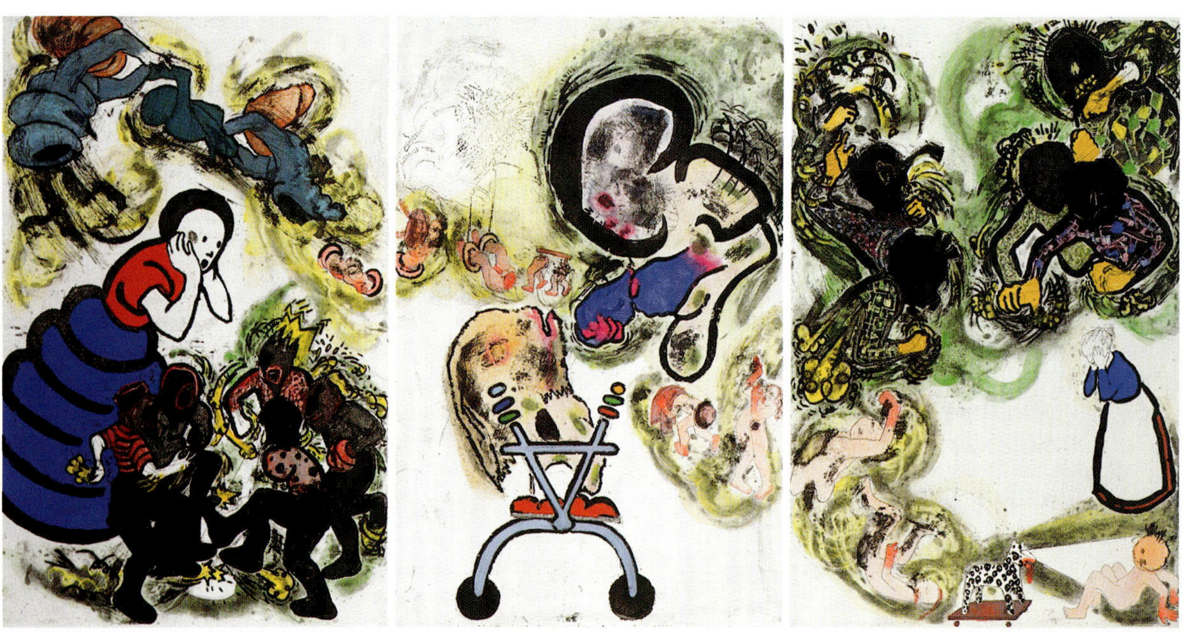

Vladimir Filipenko (b.1945)   RE798 *Sad Clown*   Woodcut 612 x 770 mm

Oona Grimes (b.1957)   RE803 *The Big Monster Secret* 1996   Etching 595 x 1185 mm

General Index

Index of People

Fellows, Associates and Honorary Fellows of The Royal Society of Painter-Printmakers are set in bold;
page references to plates are in bold italics.